Functional Foods for Disease Prevention II

Medicinal Plants and Other Foods

ACS SYMPOSIUM SERIES **702**

Functional Foods for Disease Prevention II

Medicinal Plants and Other Foods

Takayuki Shibamoto, EDITOR
University of California at Davis

Junji Terao, EDITOR
University of Tokushima

Toshihiko Osawa, EDITOR
Nagoya University

Developed from a symposium sponsored by the Division
of Agricultural and Food Chemistry at the 213th National Meeting
of the American Chemical Society,
San Francisco, California,
April 13–17, 1997

American Chemical Society, Washington, DC

Library of Congress Cataloging-in-Publication Data

Functional foods for disease prevention / Takayuki Shibamoto, Junji Tarao, Toshihiko Osawa, editors.

p.   cm.—(ACS symposium series, ISSN 0097–6156; 701–702)

"Developed from a symposium sponsored by the Division of Agricultural and Food Chemistry at the 213th National Meeting of the American Chemical Society, San Francisco, California, April 13–17, 1997."

Includes bibliographical references and indexes.

Contents: I. Fruits, vegetables, and teas — II. Medicinal plants and other foods.

ISBN 0–8412–3572–4 (v. 1). — ISBN 0-8412–3573–2 (v. 2)

1. Medicinal plants—Congresses. 2. Functional foods—Congresses.

I. Shibamoto, Takayuki. II. Terao, Junji. III. Osawa, Toshihiko. IV. American Chemical Society. Division of Agricultural and Food Chemistry. V. American Chemical Society. Meeting (213th : 1997 : San Francisco, Calif.) VI. Series.

RS164.F87   1998
615′.32—dc21   98–6978
                                                                CIP

Foreword

THE ACS SYMPOSIUM SERIES was first published in 1974 to provide a mechanism for publishing symposia quickly in book form. The purpose of the series is to publish timely, comprehensive books developed from ACS sponsored symposia based on current scientific research. Occasionally, books are developed from symposia sponsored by other organizations when the topic is of keen interest to the chemistry audience.

Before agreeing to publish a book, the proposed table of contents is reviewed for appropriate and comprehensive coverage and for interest to the audience. Some papers may be excluded in order to better focus the book; others may be added to provide comprehensiveness. When appropriate, overview or introductory chapters are added. Drafts of chapters are peer-reviewed prior to final acceptance or rejection, and manuscripts are prepared in camera-ready format.

As a rule, only original research papers and original review papers are included in the volumes. Verbatim reproductions of previously published papers are not accepted.

ACS BOOKS DEPARTMENT

Contents

INDEXES

Preface

"FUNCTIONAL FOODS" are called by many different names, including designer foods, pharmafoods, nutraceuticals, medical foods, and a host of other names, depending on one's background and perspective. Recently, food components that possess such biological characteristics, such as anticarcinogenicity, antimutagenicity, antioxidative activity, and antiaging activity, have received much attention from food and nutrition scientists as a third functional component of foods, after nutrients and flavor compounds. This symposium focuses on the latest scientific research and the impact of this research on policy and regulation of functional foods. A major objective of the symposium was to provide a forum for interaction among food chemists, nutritionists, medical doctors, students, policy makers, and interested personnel from industries.

The two volumes of Functional Foods: Overview and Disease Prevention cover the most recent research results and state-of-the art research methodology in the field. In addition, current perspectives on functional foods and regulatory issues are also presented. The contributors are experts in the area of functional foods and were selected from many countries, including the United States, Canada, the Netherlands, Germany, Finland, Israel, Japan, Korea, Thailand, India, and Taiwan.

This book provides valuable information and useful research tools for diverse areas of scientists, including biologists, biochemists, chemists, medical doctors, pharmacologists, nutritionists, and food scientists, from academic institutions, governmental institutions, and private industries.

This volume contains perspectives associated with the current status of functional foods in the United States, Europe, and Japan. Medicinal plants, such as garlic and rosemary, are introduced. Biological activities of marine and processed foods, including antioxidative, anticarcinogenic, antimutagenic, and cholesterol-lowering, provided by marine products and processed foods are outlined. In addition, determination methods for various functional principles in foods are reported.

Acknowledgments

We appreciate contributors of this book and participants of the Symposium very much. Without their effort and support, the Symposium would not be successful. We thank Tomoko Shibamoto, Sangeeta Patel, and Takashi Miyake for their assistance in organizing and proceeding the Symposium. We are indebted to Hiromoto Ochi who donated a fund to award 10 outstanding papers.

We acknowledge the financial support of the following sponsors: AIM International Inc.; Amway Japan Ltd.; Avron Resources Inc.; Fuji Oil Co., Ltd.; Green Foods Corporation; Kikkoman Corporation; Mercian Corporation; Mitsui Norin Co., Ltd.; Morinage & Co., Ltd.; Nikken Fine Chemical Co., Ltd.; Nikken Foods Co., Ltd.; Taiho Pharmaceutical Co., Ltd.; The Calpis Food Industry Co., Ltd.; The Rehnborg Center for Nutrition & Wellness and Nutrilite, Amway Corp.; UOP; Yamanouchi Pharmaceutical Co., Ltd.; and The Division of Agricultural and Food Chemistry of the American Chemical Society.

TAKAYUKI SHIBAMOTO
University of California at Davis
Department of Environmental Toxicology
Davis, CA 95616

JUNJI TERAO
Department of Nutrition, School of Medicine
University of Tokushima
Tokushima 305, Japan

TOSHIHIKO OSAWA
Department of Applied Biological Sciences
Faculty of Agriculture
Nagoya University
Chikusa, Nagoya 464-01, Japan

Contents (Volume I)

Tea and Related Compounds

INDEXES

PERSPECTIVE AND OVERVIEW

Chapter 1

Recent Progress of Functional Food Research in Japan

Toshihiko Osawa

Department of Applied Biological Sciences, Nagoya University, Chikusa, Nagoya 464-01, Japan

"Functional Foods" research project started first in 1984, under the sponsorship of the Japanese Ministry of Education, Science and Culture, and the concept for "tertiary function" has been proposed first.   This paper mainly focused on foods with "tertiary function" which are expected to contribute   to disease prevention by modulating modulation of physiological systems such as immune, endocrne, nervous, circulatory and digestive systems. Following the activation and development of systematic and large-scale "Functional Foods Research" sponsored by the Ministry of Education, Science andCulture, a national policy has approved "Functional Foods" in terms of FOSHU (Foods for Specified Health Uses) by Minstry of Welfare.   Until now, more than 100 food items have been approved and introduced to market as FOSHU according to new legislation by minstry of Welfare.   In this paper, I explain the reason why "Functional Foods"started in Japan, and I also discuss the background and recent progress of Functional Food Research in Japan.   In this paper, we also review the recent progress of research on functions of dietary antioxidants as one example for "Functional Foods".

Definition of Functional Foods

Recently, many research have been carried out on food components which have "tertiary" function.   The primary or nutritional function is the fundamental and basic function and much public interest has been focused on "primary function" when the most Japanese suffered extreme food shortage.   During the improvement in life-style in the 1960's, Japanese graduary put more focus on sensory satisfaction along with the rapid development of industry.   By the 1980's, much attention has been focused on the foods having "tertiary" functional activity.   With the definition of functional activity, "primary" function means the role of standard nutrient components, and "secondary" function has been difinited as "sensory" functions related with flavour, taste, color and texture etc.   By the 1980's many Japanese Scientists began to recognize the importance of the concept of prevention of age-related and gerriatric disease through daily dietary habits.   As shown in Table I,   recent research had shown that a variety of food components could be expected to disease prevention by modulating physiological systems such as immune, endocrne, nervous and circulatory and digestive systems.   In 1984, Japanese Ministry of Education, Science and Culture funded basic scientific research at Universities to design and create the physiologically functional foods (simply, functional   foods) on the basis of substances which have tertiary functions.

Table I.   Functions of Food

Primary function	nutrition
Secondary function	sensory satisfaction
Tertiary function	modulation of physiological systems (immune, endocrne, nervous, circulatory and digestive)

Recent Development of Functional Foods Research

In 1992, a new academic research project (Grant-in Aid for Scientific Research on Priority Area No.320) has been created under the sponsorship by the Japanese Ministry of Education, Science and Culture.   This third 3-year project entitled "Analysis and Molecular Design of Functional Foods" has been chaired by Prof. S.Arai (1).   Total of 59 research teams from 23 Universities joined together and details of the research program are shown in Table II.

4

Table II. Three Year "Grant-in-Aid" Research Program for "Functional Foods" Headed byDr. S. Arai

I. Analysis and Design of Body-regurating Factors of Foods
  a. Factors of the active form
    1. To fanction by inducing endogenous substances
    2. To function as if they were endogenous substances
  b. Factors in the precursor form (Protein-derived peptides)
    1. To function at the preabsorptive stage.
    2. To function at the postaborptive stage.
II. Analysis and Design of Body-defending factors of Foods
  a. Factors involved in the immunological Mechanism
    1. Immunostimulants
    2. Immunosuppressants
  b. Factors involved in the immunological mechanism
    1. Anti-infection factors
    2. Anti-tumor factors
III. Development of a Technological Basis for the Design of Functional Foods
  a. Design of microscopic structures
    1. Molecular breeding
    2. Molecular tailoring
  b. Design of Macroscopic Structures
    1. Development of new methods for the structure conversion
    2. Development of new methods for structural analysis

The design and construction of functional foods would be positioned to use for targeting to enrich the concentration of the quantities of functional food components, or to remove some unfavorable toxic components. In 1990, "Designer Food " project started in U.S.A., and new concept "Food Phytochemicals for Cancer Prevention" has been created (2). In 1995, International Conference entitled "Food Factors for Cancer Prevention" has been organized in Hamamatsu in December 1995, and more than 1000 participants attended this Conference. The conference brought together leading researcher from all over ther world to present the most up-to-date findings on the role of diets in cancer prevention. The proceedings of this Conference has been published in 1997, and contain more than 100 papers (3), and new concept "Food Factors"is now acceptable not only in Japan but also U.S.A. and Europe. Next International Conference for food factors is scheduled in Kyoto in December 1999.

Definition of Food for Specific Health Use (FOSHU)

In the late 1980's the Japanese Ministry of Health and Welfare has started to make an investigation to establish a category of foods which have health promoting effects to reduce the escalating cost of health care in Japan, because of the decreased consumption of fruits and vegetables by the emerging Western trends in Japanese diets.

From these backgrounds, the Japanese Ministry of health and Welfere decided to introduce new concept "Food for Specific Health Use" (FOSHU). The Japanese Ministry of Health and Welfare defined FOSHU as "processed foods containing ingredients that aid specific bodily functions in addition of nutritions" which:

O to which some ingredients to help get into shape are added;
O from which allergens are removed;
O the result of such addition/removal is scientifically evaluated;and to which the ministry of Health and Welfare has given permission to indicate the nature of effectiveness to the health.

In order to get an approval for FOSHU, each food product should be judged based on the comprehensive examination and permission/approval will be issued by MHW only to those food products that have cleared the examination of product-specific documents submitted by the applicants. Such conprehensive examination is conducted by the experienced specialists in a broad research area.. Until now, more than 100 food items have been approved and introduced to market as FOSHU according to new legislation by the Minstry of Health and Welfare (MHW). In order to get permission/approval, the applicants are required to use an ingredient that is approved, then they must ask MHW for the examination of the FOSHU product that they have developed using that ingredient (4). Now, there are 11 categories of the ingredients for FOSHU products:

1)Dietary fiber 2)Oligosaccharides 3)Sugar alcohols 4)Polyunsaturated fatty acids 5)Peptides and proteins 6)Glycosides, isoprenoids and vitamins 7)Alcohols and phenols 8)Cholines 9)Lactic acid bacteria 10)Minerals 11)Others.

After 10 years of development of FOSHU products, we are greatly disappointed and have to clarify the distinction between "Functional Foods" and FOSHU, because MHW decided not to use the term "Functional Foods" for the legislation. Many researchers in academic mainly put the focus on prevention of life-style related diseases including cancer, heart diseases and cerebravascular diseases etc by food components which have tertiary functions including new physiological activities. Now, the term "Functional Foods" became much popular and evaluated as the best term compared with "Designer Foods" and "Nutraceuticals". Our research group has been involved in development of novel type of dietary antioxidants for protection in oxidative stress. In this paper, we also review the recent progress of research on functions of dietary antioxidants as one example for "Functional Foods".

6

Case of Functional Foods for Prevention of Oxidative Stress

Although "tertiary function" means many different type of biological activities, much attention has been focused on prevention of life-style related diseases, in particular prevention of diseases induced by oxidative stress (5).   Our research group has been involved in isolation and identification of dietary antioxidants for a long time, and now developing novel and convenient evaluation systems by application of immunochemical methods for prevention of life-style related diseases such as atherosclerosis, diabetes and cancer etc.

Oxidative Stress and Diseases.   Recently, much attention has been focused on oxidative damages caused by degradation products or free radicals formed during lipid peroxidation of the cell membranes, although the active species were not identified. Excess production of oxygen radical species such as hydroxy radicals can easily initiate the lipid peroxidation in the cell membranes to form the lipid peroxides. Lipid peroxidation is known to be a free radical chain reaction which takes place in *in vivo* and *in vitro* and forms lipid hydroperoxides and secondary products such as MDA and 4-hydroxynonenal (HNE).   These lipid peroxides are highly reactive and have been shown to interact with many biological components such as proteins, amino acids, amines, and DNA.   Until now, many different type of detection and quantification methods of lipid peroxides have been developed by an instrumental analyses including application of HPLC.   Recently we have been involved in developing   novel type of evaluation systems for oxidative stress using immunochemical methods by application of polyclonal antibodies which are specific to 13-hydroperoxy linoleic acid (13-HPODE) (6), malondialdehyde (MDA) and 4-hydeoxynonenal (HNE) (7), because immunochemical methods are specific, simple and convenient.   We have also succeeded in developing a novel monoclonal antibody which are specific to HNE-modified BSA (8).   There are many indications that lipid peroxidation may play an important role in many age-related diseases including carcinogenesis, and there is speculation that oxidative damage can occur in DNA during the peroxidative breakdown of the membrane polyunsaturated fatty acids, in particular, oxidation of 2'-deoxyguanosine to 8-hydroxydeoxy-guanosine (8-OH-dG) by hydroxy radical. Although development of many sensitive methods for the detection of 8-OH-dG, in particular, ECD (Electrochemical Detector) equipped HPLC techniqui became the most popular method, despite the cost of the apparatus and requirement of many steps for sample preparation (9).   In order to develop a more sensitive and convenient method, we decided to make an evaluation for the protective role of dietary antioxidants against to oxidative damages by monitoring the amount of 8-OH-dG in biological samples using the monoclonal antibody method (10).   After obtaining specific monoclonal antibody to 8-OH-dG, we have been we develope a new ELISA (enzyme-linked immunosorbent assay) method in quantitating 8-OH-dG by competitive inhibition (11).   These immunochemical methods for detection of lipid peroxidation products and oxidatively damaged DNA are very specific and useful technique to investigate the lipid peroxidation mechanism from the viewpoint of molecular level.   By application of the immunochemical technique to the antioxidative assay systems, we can make the reliable, simple and convenient evaluation methods of antioxidative substances.

Dietary Antioxidants as Functional Foods. From our hypothesis that endogenous antioxidants in plants must play an important role for antioxidative defense systems from oxidative stress, an intensive search for novel type of natural antioxidants has been carried out from numerous plant materials, including those used as foods, and we have isolated and identified a number of lipid-soluble and water-soluble dietary antioxidants from crop seeds, sesame seeds and some spices.

Rice (*Oryza sativa* Linn.) is the principle cereal food in Asia and the staple food nearly half of the world's population. Recently, the authors succeeded in isolating and identifying an antioxidative component present in rice hull as isovitexin, a C-glycosyl flavonoid. The authors have also been involved in isolation and identification of antioxidative pigments from black rices, because black rice seeds have an ability to maintain the viability even after the long-term storage, although the white rice seeds lost viability on shorter storage. By the large scale purification and isolation of the antioxidative pigments, cyanidine-3-O-β-D-glucoside (C3G) was isolated and identified, which was found to exhibit the strong antioxidative activity in the acidic regions. C3G was also isolated and identified as the antioxidative pigments in the black and red beans, and relationship between the variety and contents of these antioxidative pigments and also antioxidative mechanism of C3G have been examined (12). By our detailed examination on antioxidative mechanism of C3G, C3G was found to have the strong antioxidative activity and be converted to protocatechuic acid which also posess the antioxidative activity when scavenging free radicals (13).

Sesame seeds and sesame oils have been used traditionally in Japan, China, Korea, and also in other eastern countries. As the lipid soluble antioxidant, sesaminol, has been found to be produced in the sesame salad oil (unroasted sesame oil) during the refining process.We have also made a large scale investigation on water-soluble antioxidants, sesaminolglucosides and found four different type of pinoresinol glucosides. Although these lignan glucosides are unique in sesame seeds, we focused mainly sesaminol glucosides, which we found in sesame seeds in a large quantity (14). Sesaminol triglucoside, main water-soluble antioxidant present in sesame seeds together with mono and diglucosides (about 1% concentration). By feeding these lignan glucosides to rats, β-glucosidase of intestinal bacteria was found to be able to hydralize enzymatically sesaminol glucosides to sesaminol and glucose. Therefore, sesaminol can be available from two sources; one route is an intermolecular transformation from sesamolin during the refining process of sesame oil, and other route is an enzymatical hydrolysis by β-glucosidase of intestinal bacteria.

Sesaminol is an unique antioxidant because it has a superior heat stability, and also is able to effectively increase the availability of tocopherols in biological systems. Recently, the protective role of sesaminol against oxidative damage of low density lipoprotein has also been examined, and this data showed that the presence of sesaminol effectively reduced oxidative modificatrion of apo protein by caused by lipid peroxidation products in the presence of a peroxidation initiator. The authors found that sesaminol inhibited the oxidative modification of DNA by monitoring the excretion of 8-OH-dG using ELISA by application of monoclonal antibody in urine of rats after feeding carbontetrachloride (CCl_4) as the *in vivo* sytem (5). By this experiment, we observed a marked increase of lipid peroxidation in blood plasma induced by CCl_4 administration, and we found that sesaminol inhibited effectively the formation of TBARS, and was effective in reducing the excretion of 8-OH-dG. These data indicated the potentiality of sesaminol for inhibition of the oxidative damages in DNA caused *in vivo* system, and long term feeding experiments of sesaminol and sesaminol glucoside are now undergoing.

CONCLUSION

The term of "Functional Foods" and its concept are now internationally accepted. It seems that such an international interest in "Functional Foods" reflects the current research activities in Japan, however, MHW decided not to use the term "Functional Foods"and adopted the term FOSHU instead. Looking back on the FOSHU systems over five years after introduction to Japanese Food Industries, many companies have a frustration because of the limitation of health claim and labelling. However, research scientists in both academics and industries in many countries including Japan are now ready to develop a new age "Functional Foods" which are evaluated by clinical testing and approved to make a health labelling for prevention of life-style related diseases including atherosclerosis, diabetes and cancer etc.

There are many indications that lipid peroxidation plays an important role in carcinogenesis, although there is no definite evidence. Modification of immune systems are also very important approach to prevent diseases, however, I mainly focused on prevention of oxidative stress by dietary antioxidative components as one of the examples for "Functional Foods" in this paper. Our group started novel approach to develop evaluation and detection methods of lipid peroxidation products by application of monoclonal and polyclonal antibodies. We have also started our new project to investigate a novel biomarker of oxidative DNA damage other than 8-OH-dG. Of course, research efforts on metabolic pathways of dietary antioxidants in the digestive tracts are also required. Although these approach started just recently, they help us to understand the relationship between antioxidative activity and cancer prevention.

REFERENCES

1. Arai, S. (1996) Studies on Functional Foods in Japan-State of the Art, Biosci. Biotech. Biochem., 60, 9-15.
2. Huang, M-T., Ho, C-T., Osawa, T. and Rosen, R.T. (1994) Food Phytochemicals for CancerPrevention I and II, ACS, Washington.
3. Ohigashi, H., Osawa, T., Terao, J., Watanabe, S. and Yoshikawa, T. eds. (1997) Food Factors for Cancer Prevention, Springer, Tokyo.
4. The Japan Health Food and Nutrition Food Association (1995) A Quick Guide to: Food for Specified Health Use.

5. Osawa,T.,Yoshida,A., Kawakishi,S., K. Yamashita,K. and Ochi,H. (1995) Protective Role of Dietary Antioxidants in Oxidative Stree, Oxydative Stress and Aging, R.G. Cutler, J. Bertman, L. Packer and A. Mori, eds.,p.367-377, Birkhauser Verlag Basel/Switzerland.

6. Kato,Y.,Makino, Y. and Osawa,T. (1997) Characterization of a Specific Polyclonal Antibody against 13-Hydroperoxyoctadecadienoic acid-modified Protein, J. Lipid Res., 38, 72-84.

7. Uchida, K., Toyokuni, S., Nishikawa, K., Kawakishi, S., Oda, H., Hiai, H. and Stadtman, E.R. (1994) Michael Addition Type 4-Hydroxy-2-nonenal Adducts in Modified Low Density Lipoproteins: Markers for Atherosclerosis, Biochemistry 33, 12487-12494.

8. Toyokuni, S., Miyake, N., Hiai, H.,Hagiwara, M., Kawakishi, S., Osawa, T. and Uchida, K. (1995) The Monoclonal Antibody Specific for the 4-Hydroxy-2-nonenal Histidine Adduct, FEBS Lett. 359, 189-191.

9. Toyokuni,S.,Tanaka,T., Hattori,Y., Nishiyama,Y. Yoshida,A., Uchida,K., Ochi, H. and Osawa,T.(1996) Quantitative Immuno-histochemical Determination of 8-hydroxy-2'-Deoxyguanosine by a Monoclonal Antibody N45.1: Its Application to Ferric Nitrilotriacetate-induced Renal Carcinogenesis Model, Laboratory Invest., 76(3), 365-374.

10. Hattori,Y.,Nishigari,C.,Tanaka,T.,Uchida,K.,Nikaido,O.,Osawa,T.,Hiai,H., Imamura,S.and Toyokuni,S.(1996) Formation of 8-Hydroxy-2'-deoxy-guanosine in Epidermal Cells of Hairless Mice after Chronic UVB Exposure, J. Invest. Dermat., 107(5), 733-737.

11. Erhola,M.,Toyokuni,S.,Okada,K.,Tanaka,T.,Hiai,H.,Ochi,H.,Uchida,K.,Osawa, T.,Nieminen, M.M.,Alho,H. and K-Lehtinen, P. (1997) Biomarker Evidence of DNA Oxidation in Lung Cancer Patients: Association of Urinary 8-Hydroxy-2'-deoxyguanosine Excrerin with Radiotherapy, Chemotherapy, and Response to Treatment, FEBS Letters, 409, 287-291.

12. Tsuda, T., Shiga, K., Ohshima, K. and Osawa T.(1996) Inhibition of Lipid Peroxidation and Active Oxygen Radical Scavenging Effect of Anthocyanin Pigments Isolated from *Phaselous vulgaris* L., Biochem. Pharmacol., 52, 1033-1039.

13. Tsuda,T.,Oshima,K.,Kawakishi,S., and Osawa, T.(1997) Oxidation Products of Cyanidin 3-O-β-D-glucoside with Free Radical Initiator, Lipids, 31, 1259-1263.

14. Katsuzaki, H., Kawakishi, S. and Osawa, T. (1994) Sesaminol Glucosides in Sesame Seeds, Phytochem., 35, 773-776.

Chapter 2

Communicating the Benefits of Functional Foods: Insights from Consumer and Health Professional Focus Groups

D. B. Schmidt[1], M. M. Morrow[1], and C. White[2]

[1]International Food Information Council, 1100 Connecticut Avenue, Suite 430, Washington, DC 20036
[2]Axiom Research Company, 2 Tyler Court, Cambridge, MA 02140

Consumer and health professional focus group research results reveal a positive climate for the acceptance of functional foods in the United States. Ten focus groups of consumers, dietitians, and family physicians were conducted in a total of five U.S. cities in 1996 and 1997. Insights from the focus groups helped to identify effective communications strategies for reaching consumers, health professionals, news media, and regulators, and noted opportunities and challenges to communicating functional foods issues to these audiences. Effective communications, based on sound science, can make the difference in fostering an environment that encourages future research for both the development of new food products and the discovery of beneficial properties of existing foods and food components.

Media and consumer interest in the functional foods trend appears to be outpacing science and regulatory policy development; communicating about functional foods is therefore a critical element in realizing this trend's potential public health benefits. The definition of functional foods is still evolving, but they are generally considered to be foods that may provide health benefits beyond basic nutrition.
    The International Food Information Council (IFIC) commissioned Axiom Research Company to conduct focus group research on functional foods in 1996 and 1997 with three different audiences: consumers, dietitians, and family physicians. The goal of the research was to assess awareness and acceptance of functional food concepts, gauge reaction to current and potential sources of functional components, and evaluate reactions to terminology.

Focus Groups as a Research Methodology

Focus groups allow researchers to create a unique setting where they can gather a wealth of in-depth information and feedback from a specific target population (e.g. consumers, voters, women, etc.) on a particular issue or subject. Typically groups last one and one-half to two hours and are comprised of eight to ten participants. Groups are carefully designed to facilitate a semi-structured interactive discussion among the participants led by a professional moderator. The open-ended format allows the moderator to probe in-depth on a wide range of questions related to the research topic. In addition, the group setting produces an interactive dynamic that stimulates participants and encourages them to react and respond to one another. Results can provide researchers with keen insights about the subject matter, and can often lead them to pursue avenues of investigation or inquiry that had not previously been considered.

Unlike quantitative research methodologies such as telephone surveys, results from focus groups are not measured statistically, and thus findings cannot immediately be projected to the general population. While conducting multiple groups helps to uncover and identify important trends, insights or findings garnered through focus groups must still be tested through quantitative research in order to verify their accuracy.

Geographic Distribution and Composition of the Focus Groups on Functional Foods

A total of ten focus groups were conducted in five cities. Four focus groups, composed of consumers, were conducted in Boston, MA and Richmond, VA in June 1996. One focus group each of dietitians and of physicians was conducted in Bethesda, MD, Indianapolis, IN, and Los Angeles, CA during February 1997.

Consumers ranged in age from 21 to 65, with the majority falling somewhere in the middle. About half of the consumers in each group reported taking supplements although none shopped at health food stores on a regular basis; two-thirds of all participants were women. They were also selected according to their degree of health-consciousness such that one group from each city was "health active," self-described as more aware and knowledgeable on nutrition than others, and the other was "casually cognizant," having more than average awareness. "Health active" participants agreed with at least two of the three following statements while "casually cognizant" participants agreed with one:

1) I am more interested in food and what I eat than are most people.
2) I shop for healthy foods more than most people do.
3) My friends and family ask me for health and nutrition advice.

An effort was made to recruit family physicians and dietitians in each of the three cities such that each focus group contained a mix of participants who varied according to the following characteristics: years practicing (twenty percent each: less than five, fifteen to twenty-four and twenty-five or more years, and forty percent five to fifteen years), use of dietary supplements (users and non users), ethnicity and gender, among others. Despite recruitment efforts for male as well as female dietitians, most participants in the focus groups of dietitians were female, reflecting the profession's gender distribution.

A common underlying belief shared by the focus group participants, dietitians, physicians, and consumers alike was that diet plays an important role in determining good health.

Sources of nutrition information

Health professionals named scientific journals and newsletters as their best sources of nutrition information, but acknowledged that they often first hear of new research through the popular media or from patients. Later, when the journals arrive, they educate themselves on the topic.

Most physicians and dietitians think the news media are powerful information sources, and appreciate that they help raise awareness among patients about the importance of proper diet. However, the media are also viewed as a "double-edged sword" as many patients become confused by the vast amount of diet and health information, and as a result can be led astray by misinformation. Some health professionals said their patient education process consists primarily of debunking the myths their patients have accepted as truths.

Consumers also named the popular media as their number one source of nutrition information. By synthesizing information from friends and relatives, physicians, government sources and product labels, consumers are able to confirm, explain, reassure and remind themselves of what they have heard from the media. Consumers in these and previous IFIC focus groups said that news magazine shows, for example, *60 Minutes* and *Dateline*, are credible sources of food and health information and these sources are appreciated for both their depth and entertainment value.

Functional Foods Awareness

Consumer and health professional awareness of the *concept* of functional foods is high. A surprising number of consumer participants were able to list foods and their specific components that have health benefits associated with them. Top of mind foods and their beneficial components among consumers include milk (calcium) and fruits and vegetables (antioxidants). Fiber-fortified cereals were less readily mentioned by consumers but were top of mind for health professionals. Consumers associate calcium with osteoporosis prevention, antioxidants with slowing the process of aging, and fiber with prevention of colon cancer. However, regardless of these associations, there appears to be confusion among consumers as to technically what is or is not considered a functional food. Reflecting this lack of clarity, some consumers mentioned issues related to fats, oils, and calories which had not been intended for discussion.

Terminology

One of the goals of the research was to determine a universal term for these types of foods that would be appealing to both consumers and health professionals. By far, consumers preferred the term "functional foods," over other alternatives such as "designer foods," "pharmafoods," "nutraceuticals," or "phytochemicals." One consumer explained, "To me functional means there is a purpose other than just its nutritional value." Dietitians reluctantly selected "functional foods" as a general term, and physician opinion was divided with a slight majority preferring "optimum food." Regardless of their first choice, "nutraceuticals" received the least support from all the groups.

While classifying foods with functional characteristics may be important to food manufacturers and others within the food industry, it is questionable whether such a term is actually necessary for consumer acceptance of individual products which could stand on their own perceived attributes.

Functional Foods Attitudes

In an attempt to avoid placing too much emphasis on "magic bullets," health professionals indicated that they would prefer to recommend functional foods to their patients for their overall <u>nutritional</u> value rather than for any benefit attributed to an individual component. However, many said they would also mention the added health "bonus" of eating the particular food.

Consumers were receptive to learning more about functional foods, and would be willing to incorporate these foods into their diets. With a few exceptions, they were particularly favorable toward foods in which the functional components occur naturally. Younger consumers are less concerned about potential long-term benefits of functional foods, like cancer prevention, and more interested in short-term energy boosts from sports bars, for example. Aging consumers on the other hand are very interested in eating foods with long-term health benefits.

Alternative Sources and Quantifying Components

While all groups reportedly prefer traditional sources of functional components, they also see practical advantages to having alternative sources of the same functional food component. One dietitian acknowledged, "You have to have a variety on the market, and then you can make recommendations based on your patient's food preferences and lifestyle."

Fortification and quantities of particular components needed for desired effects were of interest to participants because they liked the idea of reaping the potential benefits of a component without having to consume large quantities of the food containing it. However, all parties expressed a desire for guidelines recommending daily intake quantities, reflecting a concern about over-consumption.

Focus group participants were asked to react to a variety of hypothetical generic products that may or may not be found in the marketplace. Most participants felt positive about the use of biotechnology in food, however a few felt that any enhancement through biotechnology would be "altering nature". Two types of possibilities included hybrid plants enhanced through biotechnology to produce desired components they wouldn't ordinarily contain (i.e. lycopene in apples and potatoes), and original sources "boosted" to contain twice the amount of the desirable component (tomatoes enhanced with lycopene). Another possibility mentioned in all focus groups was that of enhanced yogurt (with twice as many live cultures), although the scientific data on the effect cultures have on digestion is still unclear. Other products discussed that hypothetically could be fortified with live cultures were cereal, milk, and water. The latter, fortified water, was not well received by participants.

Fortified snack foods, already perceived to be relatively high in calories, were products that received mixed reviews from the ranks. Those in favor of fortified snacks liked the idea of getting a "bonus" from a food that was likely to be eaten regardless of health benefits. Yet other consumers immediately jumped to concerns about fat content of such products. Objection from some physicians was on the grounds that recommending these products would be tantamount to endorsing a poor diet:

"We're also trying to do something that has a positive impact to change behavior ultimately. Isn't that what we're supposed to be doing? We're not just supposed to be adding elements to food and skirting the real issue, which is changing behavior--having people become more healthful in their approach to what they eat."

Physicians, dietitians and consumers are not overly enthusiastic about a "functional food pill." In general, health professionals feel that under very specific conditions—the component was truly helpful, the pill provided a "healthy dose," and the patient could not or would not easily consume it in its natural source—they might recommend such pills. Concern focused on a worry about an evolving culture of "pill poppers," and people seeking easy solutions instead of making healthful lifestyle changes. Consumers generally said they would accept the pill form of functional food components only if their physicians recommended that they get more of the component in their diets.

Despite a willingness to consider alternate sources of functional food components, health professionals agreed that they would always prefer to see their patients eating the traditional source as part of a balanced diet: "Personally, I prefer the food form because there are other things in the food that you won't ever get out of a capsule. There's more than just lycopene in tomatoes."

Confidence in food product claims and labeling

Health professionals were more familiar than consumers with government regulations of food and supplements. "In my opinion, food product claims are more believable because of the way they're regulated," said one health professional. "Supplements aren't well regulated, but food products cannot make claims without approval of the FDA. They're really strict about how that's regulated." On the other hand, most consumers erroneously thought supplements were more strictly regulated than food, and believed supplement manufacturers were either more, or equally, as credible as food manufacturers. Further probing indicated that consumers equate supplements with drugs and assume they are regulated similarly.

Consumer skepticism of product labeling distinguished between what the front of the label said, viewed as advertising, versus what was believed to be the more credible Nutrition Facts panel printed on the side or back of the packaging.

An example of label wording for a functional food that would be acceptable to participants might be, "This product contains lycopene, which *may* reduce risk of some cancers" rather than a bolder statement, "This product prevents cancer."

IFIC's focus groups indicate that there could be a strong future for functional foods. Consumers expect to see functional food products in their supermarkets alongside traditional products. While the functional foods trend offers growth opportunities for the conventional food, natural food and supplement industries, public health goals will most likely only be met if the mainstream food industry leads the way in product innovation and marketing.

Educational impact of the focus groups

By the end of each session, many focus group participants expressed great interest in somehow incorporating functional foods, especially those discussed, into their diets. The consumers in the Richmond focus groups were particularly affected, stating that they planned on incorporating more of functional foods discussed (especially tomatoes) into their shopping and eating patterns. The Boston focus group participants were initially more skeptical, but they still indicated that they, like those in Richmond, would try to both buy and eat more functional foods. Nearly all said they would give more thought to, and feel better about, eating the functional foods which had been discussed. "Yes, I would, I would eat more of the yogurt that I buy," confirmed one consumer. A physician acknowledged that, "Good medical schools have programs now; we cannot shut our minds to alternatives to medicine."

The overwhelmingly positive response of participants to the focus group discussions suggests that the public is ready to be educated and is receptive to receiving information on functional foods, as long as it is backed by credible sources and sound science.

Current Communications Environment for Functional Foods

The current communications environment for functional foods is comprised of a complicated blend of popular news media, consumers, and information provided by health professionals, the food industry, and regulators. Functional food is still a new, hot topic in the media. IFIC media monitoring found that functional foods related articles in popular magazines comprised six to thirteen percent of articles on diet and health each month from January to April of 1997.

Consumers, as evidenced by the focus groups, are extremely receptive to learning more about functional foods. However, health professionals do not always recognize their patients' receptivity. Doctors said they often feel pressure from patients to prescribe a pill in response to an illness; consequently they are somewhat reluctant to offer advice which might seem more like a home remedy than medicine. Dietitians in general may need a paradigm shift to view foods as having properties beyond their basic nutritive value.

Meanwhile, the food industry harbors two schools of thought. The first being that in order for functional foods to be worth pursuing, they need approved health claims which would require the investment of a considerable amount of time and money. An alternative way of thinking is that health claims on the packages are not necessary due to general awareness of the benefits of particular components. Consequently, content labeling without product health claims could be sufficient.

Regulators have been quiet as to their position on functional foods. The official position of the Food and Drug Administration is that there is no need to create a new category of food to accommodate functional foods; all food can fit into existing categories as food, supplements, medical foods, or foods for special dietary use.

Growing Markets

Rising awareness of functional foods and consumer demand should lead to growth in conventional foods, supplements, and natural foods, rather than just competition between markets. In order to have a major public health impact, the food industry and nutrition community should lead the way for functional foods so that mainstream recognition is achieved for the category.

Communications Opportunities for Functional Foods

Opportunities abound for communicating information about functional foods. Consumers' minds are open to new information because they are just beginning to learn about the topic and have yet to solidify opinions. Likewise, now is the time to inform opinion leaders on functional foods issues.

The science behind functional foods is in a dynamic era of research, as indicated by the number of studies presented at the American Chemical Society's *Symposium on Functional Foods*. The proliferation of studies offers the media a vast array of stories to tell, which they have certainly done thus far.

Information on functional foods is well-received by the public for two reasons in particular. As the Baby Boom generation ages and feels those accompanying "aches and pains" it becomes more interested in healthful products. Also, consumers want *positive* information. They prefer to hear what is healthful to eat rather than what to eschew.

Communications Challenges for Functional Foods

Communications challenges are varied. For starters, consumers are easily confused by the glut of popular information on food issues and are unsure how to respond to contradictory or changing recommendations. Lack of recognition by the FDA along with varied terminology for functional foods is associated with a category identity crisis. Definitions for functional foods vary in scope; what one person calls a functional food may be considered nothing more than nutritious by another. Furthermore, health professionals tend to be caught in a nutrition paradigm where foods are but the building blocks of a balanced diet and medications the answer to illness.

While the news media can be an excellent source of information regarding functional foods, they may do a disservice when painting functional foods in a very "hi-tech" light. Consumers prefer traditional sources of functional food components and are turned off by laboratory images. Although functional foods have not been opposed by regulators, the down side to their quiet position is a lack of endorsement statements quotable by media and industry sources.

Strategies for Communications on Functional Foods

The following strategies for communications on functional foods take into consideration the aforementioned opportunities and challenges:

1) Educate opinion leaders and health professionals to match consumer readiness.

2) Promote the need to identify credible, scientific criteria for defining functional foods.

3) Place new research findings in the context of related studies to avoid conclusions drawn from single studies.

4) For maximum acceptance, place functional foods in the most traditional food context possible, avoiding images of beakers and microscopes.

5) Make all communication regarding functional foods reasonable, responsible information.

6) In absence of government regulation, pay extra effort to backing up claims with sound science.

7) Use terminology like "may reduce risk" rather than "will prevent," to describe the attributes of components, as the latter statement is equated with drugs by the FDA.

8) Do not wait for approval of specific health claims; list components on labels and consumers will respond.

9) Look for ways to incorporate functional foods into existing public health campaigns. For example, promote functional foods as a way to meet the "5-A-Day" goal established by the National Cancer Institute to motivate Americans to consume more fruits and vegetables.

10) Finally, and most importantly, emphasize maximum consumption of functional components through a varied diet.

Chapter 3

Developing Claims for Functional Foods

R. E. Litov

NutraTec, 3210 Arrowhead Drive, Evansville, IN 47720–2504

There are no forthcoming regulations for functional foods in the U.S.A. However, there are three regulatory categories within which they may be currently marketed: 1.) food, 2.) dietary supplement, and 3.) medical food. Food "health claims" are the most restrictive health messages, requiring the greatest commitment of time and resources for a new claim approval. Dietary supplement "nutritional support statements" have the most flexible language, but are the least persuasive. Medical food "dietary management claims" are the most compelling, but the products have limited application and need significant clinical research support. Although information on food labels does influence the buying habits of surveyed consumers, few use the health claims on food labels and most are skeptical about the health claims made. Reports on nutrition and health studies confuse consumers because of conflicting results. Therefore, to maximize the effectiveness of health messages for functional foods, they need to be 1.) clearly communicated to consumers and 2.) credible, ie., supported by adequate research that convinces consumers of the link between the bioactive agent and the health benefit.

The use of food for good health has been practiced since early humankind. Certain plants and their components have been used by shamans, medicine men, and witch doctors to treat or prevent a multitude of ailments. Ancient cultures of China, India, and around the Mediterranean Sea used cooling foods for treating hot diseases and warming foods for cold illnesses. This system of hot and cold foods to treat disease is still practiced today in many parts of the world. As modern science and technology advanced, food fortification was introduced in the mid-20th century. Iodine added to salt, vitamins A and D to milk, and B vitamins to flour are now common practice. In the 1990's, interest in a category of foods commonly called "functional foods" has grown.

Functional Foods

There is no universally accepted or legal definition of a functional food. A suggested general definition is as follows: "Functional food or beverage products contain safe substances, are promoted for health benefits beyond meeting nutritional needs of growth and maintenance, and are included in a diet." The additional health benefits that functional foods are being designed to provide include preventing and helping to treat disease, enhancing physical performance, and improving the quality of life.

Kellogg's All-Bran may be considered one of the first functional foods. In 1984, this breakfast cereal was the first food product that publicized the relationship between foods and disease on its label. Because of its fiber content, it was promoted as part of a diet that could reduce the risk of certain kinds of cancer. The health message had been developed in collaboration with the National Cancer Institute. A controversy followed regarding the use of health messages with food products. This prompted activity that later resulted in changes in the food regulations with the passing of the Nutrition Labeling and Education Act of 1990.

Product Claims

Product claims are valuable to food manufacturers and marketers in promoting their products. They can be used effectively to differentiate a product from the many similar products in the highly competitive food industry. Health messages associated with products can further educate consumers about the relation between food and health, an important objective by government agencies. Also, product claims can help consumers make informed decisions in the choice of their foods and nutrient intake.

Health messages for functional food products can be made depending on what ingredients and levels are used and how the product is positioned. An overview of the current regulatory categories within which such products can be positioned and the limitations and allowances for health messages are given in the next section.

Regulatory Categories for Functional Foods

It is highly unlikely that new regulations will be forthcoming to establish a separate regulatory category for functional foods in the U.S.A. However, this should not discourage marketers from developing and commercializing functional foods. The advantages to using the current regulatory guidelines for food and health care products are a.) not having to wait for new regulations and b.) revising existing regulations to accommodate specific objectives of functional foods would be easier than developing new regulations from the beginning. Presently, a new functional food product may be eligible for regulatory status as a 1.) food, 2.) dietary supplement, or 3.) medical food. Because the drug regulatory route is so cost and time prohibitive, it is not included in this discussion. The basic differences among the three regulatory categories are highlighted below and in a table. This article is not intended to be a thorough review of the regulatory requirements. Deciding which regulatory route to follow is an important early step in developing new functional food products. The choice depends on many factors, including business goals and available resources. It should be realized that there are many areas within these regulations that are subject to interpretation. Much of the present uncertainty for functional foods will resolve as the Food and Drug Administration (FDA) begins to establish a clearer pattern of enforcement and as case law emerges.

Health messages need substantiation from scientific and medical studies to establish clearly the link between the bioactive agent of a functional food and a health benefit. These studies may include epidemiologic, biological mechanism, and intervention trials. The extent of such scientific support depends on the regulatory category and language of the health message to be used. Ideally, at least two randomized, placebo-controlled, double-blind human intervention trials showing significant and clinically meaningful health benefits should be among the scientific support publications.

Food. The Nutrition Labeling and Education Act (NLEA) of 1990 (*1*) for foods allow for three types of claims: 1.) nutrient content, 2.) structure-function, and 3.) health. Nutrient content claims characterize the level of a nutrient, based on a predetermined reference amount of the food product. Claims are permitted only for those nutrients listed in the NLEA and implementing regulations, which include calories, fat, cholesterol, sodium, sugars, and those with established Reference Daily Intakes. Examples of nutrient content claims include "low in fat" and "cholesterol free." Although allowed, structure-function claims are rare or not currently used by food marketers. Health claims are used with some food products, but have several limitations including:

- The product must be safe (ie., an established nutrient or one with Generally Recognized As Safe or Food Additive status).
- The product must be a natural source of at least one of six nutrients (ie., provide at least 10% of the Daily Value for vitamin A, vitamin C, iron, calcium, protein, or fiber).
- The product must not exceed 20% of the Daily Value of risk-increasing nutrients (ie., fat, saturated fat, cholesterol, or sodium).
- There must be "significant scientific agreement" of the health benefit with respect to the totality of the publicly available scientific evidence (which is determined by the FDA).
- The model claim language is prescribed (the relationship to total diet is conveyed and no mention of product brand is allowed).

Nine health claims have been approved for foods to date. Petitions for new health claims can be submitted, but getting approval is long, difficult, and costly. Food health claims are the most restrictive health messages of any of the three regulatory categories and require the greatest commitment of time and resources for a new claim approval.

Dietary Supplement. The Dietary Supplement Health and Education Act (DSHEA) of 1994 (*2*) repositions a product as different from food, exempts it from food additive provisions, and allows for flexible product claims. A product must be labeled as a dietary supplement and cannot be represented as a meal replacement or a conventional food. Unlike foods, dietary supplements are allowed to use "nutritional support statements." There is no formal approval process, manufacturers only have to notify the FDA of the statement's use within thirty days after first marketing the product. There are four types of "nutritional support statements": 1.) classical nutrient-deficiency disease support (eg., vitamin C and scurvy), 2.) description of the intended role that the ingredient plays in affecting structure or function of the body (eg., calcium and bone structure), 3.) characterization of the mechanism by which the ingredient acts to maintain such structure or function (eg., selenium as an integral part of the active site in the enzyme glutathione peroxidase), and 4.) description of the effect on general well-being expected from consumption of the ingredient (eg., is soothing). Marketers may not make any reference to disease conditions when using these statements in labeling and advertising. In addition, the following must be prominently displayed on the label: "This statement has not been evaluated by the Food and Drug Administration. This product is not intended to diagnose, treat, cure or prevent any disease." Dietary supplement "nutritional support statements" have the most flexible language of the health messages in the three regulatory categories, but are the least persuasive.

Medical Food. Medical foods are the least formally regulated of the three categories. A definition of medical foods is as follows: "A medical food is a food which is formulated to be consumed or administered enterally under the supervision of a physician and which is intended for the specific dietary management of a disease or condition for which distinctive nutritional requirements, based on recognized scientific principles, are established by medical evaluation" (1). However, additional regulations are expected as an advance notice of proposed rulemaking was made in November of 1996 (3). Traditionally these products have been promoted through health care professionals and distributed to hospitals, nursing homes, and home health care with retail distribution in pharmacy settings. Medical food "dietary management claims" are the most compelling health messages of the three regulatory categories, but the products have limited application and need significant clinical research support.

Table I. Comparison of Health Messages Typically Used

Point of Difference	Food	Dietary Supplement	Medical Food
1. Health Message	Health Claim	Nutritional Support Claim	Dietary Management Claim
2. Approval	Preappoval	Notification	None, advanced notice of proposed rulemaking pending
3. Language	FDA prescribed	Flexible	Flexible
4. Claim Substantiation	Significant scientific agreement	Truthful and not misleading	Clinically significant

Consumer Attitudes about Food and Labels

Increasingly, consumers recognize the connection between diet and disease and many are willing to change their eating habits for better health. Food labels can inform consumers about food products and help them choose foods to realize healthier diets. Often neglected by marketers is the consumers' understanding of the health benefits of certain foods or ingredients and how they use the information given on food labels. Labeling and the health messages for products positioned as functional foods become more important in the marketing of these products to consumers. However, little research has been done in this area and only of a general nature. There is no published study of consumers' responses to specific labels and health messages of functional food products.

The American Dietetic Association (ADA) conducted a national telephone survey of 807 adults to examine their attitudes and behaviors toward nutrition and health (4). They found that 49% of consumers were very or somewhat confused by news reports on nutrition. A similar survey conducted two years later in 1997 found that 27% of consumers are confused by all the reports which give dietary advice (5). Although consumers complain that there are too many conflicting studies that confuse them, many (51%) want to hear about new nutrition studies.

The FDA conducted a national telephone survey of consumers to monitor the impact of the recent changes in food label regulations (6). When consumers were asked how they use food labels, 70% often use them to see how high or low the food is in certain nutrients. Only 33% used the label to see if something said in advertising or on the package is actually true. The part of a food label most frequently used often by consumers is the nutrient amounts, by 56% of those surveyed. The part of a food label least used often was health claims, by only 25% of consumers. When asked about the accuracy of label claims, only 31% of consumers said about all or most health claims are accurate.

Interestingly, the nutritional information provided on food labels has changed the buying habits of consumers (5). Around half of those surveyed (56%) by the ADA in 1997 switched to a different brand of the same item based on the label. In addition, 65% of consumers bought some foods more often because of the information on the label.

In conclusion, information on food labels does influence the buying habits of consumers. However, few consumers use the health claims on food labels and most are skeptical about the health claims made. This may be due in part to consumers perceiving the claims as strictly advertising and not being adequately supported by facts or reviewed by an independent authoritative organization to be credible. Also, the reports on nutrition and health studies confuse consumers because of their often conflicting results. Thus, to maximize the effectiveness of health messages for functional foods, they need to be 1.) clearly communicated to consumers and 2.) credible, ie., supported by adequate research that convinces consumers of the link between the bioactive agent and the health benefit.

Acknowledgments

The author thanks Mary Alice Springer for technical writing assistance and John J. Bushnell for reviewing the regulatory content.

Literature Cited

1. U.S. Congress. *Nutrition Labeling and Education Act of 1990.* 1990, P.L. No. 101-535.
2. U.S. Congress. *Dietary Supplement Health and Education Act of 1994.* 1994, P.L. No. 103-417.
3. Food and Drug Administration. *Federal Register.* 1996, 61(231):60661-60671.
4. American Dietetic Association. *Nutrition Trends Survey 1995.* 1995, pp 1-7.
5. American Dietetic Association. *Nutrition Trends Survey 1997.* 1997, pp 1-11.
6. Levy, A. S.; Derby, B. M. "The Impact of the NLEA on Consumers: Recent Findings from FDA's Food Label and Nutrition Tracking System," Internal Report of the Food and Drug Administration, Consumer Studies Branch, Center for Food Safety and Applied Nutrition, January 23, 1996, pp 1-6.

Chapter 4

Functional Foods

The Food Industry and Functional Foods: Some European Perspectives

O. Korver

Unilever Research Laboratory, Vlaardingen, The Netherlands

In the industrialized world, alternative lifestyles, in the way of diet, are being recommended as a preventative action to major health problems. There is a growing interest in the development of foods with properties that may have a beneficial influence on specific health states. These so called "functional foods: are currently being investigated in the food industry. When marketing these foods it is necessary to communicate their nutritional qualities. To avoid misleading the consumers, the basis of this communication must be based on sound scientific principles. In Europe, a legislation has not yet been developed. There have been many suggestions, from various groups, to define the basis for guidelines that may reflect the interests of not only the consumer, but industry as well.

Populations in the industrialized world are increasingly facing major diet-related health problems, including cardiovascular disease, obesity, diabetes, and cancer. Changes in lifestyle, of which a change in dietary habits is one, have been recommended to counter this trend. The traditional dietary recommendations of the past decades have been concentrating on the macronutrients (eat less fat, eat less saturated fat, eat more complex carbohydrates), or on whole foods (eat more fruits and vegetables).

In the past 5 years, a new trend has emerged: discussions about the introduction of so called "functional foods". The term is, in fact, an abbreviation of "physiologically functional foods". This development arose from the scientific findings that many plant components have physiological effects. Researchers started to realize that the term "non-nutritive factors", often used for these components, was only used to shield us from our ignorance about their real effects. In Asian countries, especially China, plant foods have already been used as medicine for centuries. In the West, we try to separate food and drugs. The Chinese disagree, as they say "My food is my medicine, and my medicine is my food".

Many people have attempted to define functional foods without actually coming to a clear agreement. There is, however, agreement on some boundary conditions to functional foods. When we talk about "functional foods", the following minimum conditions apply:

-Functional foods are foods, not drugs: this implies that all the quality criteria that are applied to foods, are part of the product(eg: taste and smell).
-The food, used in a realistic way, influences a physiological parameter.
-The change in the physiological parameter influences the health and performance of the individual in a positive manner.

The healthy properties of a "functional food" must be communicated to the consumer. This must be done because a consumer may decide to consume the food on the basis of the directly observable properties of the product (i.e.: whether the taste is to his/her liking), although, the "health" properties may not be directly observable from the product. The claims and messages that are given, are important for the consumer to make any decisions leading to a healthier diet.

Just as the consumer must have an understanding of the "functionality" of their foods, a producer must be given the opportunity to communicate the functionality of the food they distribute. To put restrictive regulations on making claims as to the function of the food, will possibly prevent the producer from covering the Research and Development costs. To make a claim, it must be made with ample scientific research and support, which may incur a significant cost to the producer. On the other hand, no regulations at all may make it possible to make unsubstantiated claims. This will surely lead to confusion among the consumers. Many different countries have attempted to find the "middle road", but which is the most appropriate? Before explaining some of the European developments conscerning this issue, I will bring an example from my own experience on a "functional food" avant la lettre.

At the end of the fifties the "lipid hypothesis" was postulated. This hypothesis states that dietary fat composition is an important parameter for blood lipid levels. Blood lipid levels happen to be one of the major risk factors for cardiovascular disease. There was an increased awareness in the medical circles about the "lipid hypothesis", and doctors in the Netherlands approached Unilever with the request to produce products that could be used to stimulate persons with high blood lipid levels, to change their dietary fat composition. These products, obviously, should fit into the normal dietary intake pattern.

The first product was a fat that could be used as a spread, or a frying product, and was sold in a tin via pharmacies. Later, after the introduction of the plastic margarine tub, softer table spreads could be made with the maximum quantity of unsaturated liquid oil, bound to a minimum of hard fat. On the basis of the new scientific findings, sunflower oil, with its high linoleic acid content, was used to optimize blood lipid lowering. This product still exists. It has continuously remained on the market with scientific consensus. The composition, fat level, and the communication about the product has been adapted continuously. Notably, the concept of prevention against cardiovascular disease has also been brought up. The learnings from more than 35 years of marketing this "functional food" are:

-be part of the science
-stay in line with the scientific consensus
-accept a long term development
-communicate to key opinion formers and consumers separately
-on pack claims are not the only way to communicate

This brings me back to the issue of claims and messages. When using the word "claims", it will mean all communication regarding health around the product: advertising, leaflets to KOF's, and on pack messages, etc.

Regulatory authorities and consumer associations will promote measures to prevent the misleading of consumers. Since misleading consumers will undoubtedly lead to confusion, this is also an important issue for any responsible industry. Regulators also want to avoid creating the false impression that "one product can cure". This is quite clearly a valid nutritional point, and one that should not be taken lightly.

All regulatory solutions that have been proposed, have considered two routes. The 2 main routes in coming to a solution concerning regulatory issues are: either strict legislation, or a voluntary code of conduct. The U.S. has come to a legislative solution (positive list). In Japan, there is also a law permitting Foods for Special Health Use (FOSHU), when 8 criteria are satisfactorily met. In Europe, discussions on legislation have not yet led to an agreement. This situation has led to a kind of "vacuum". Consequently, this has led to initiatives to come to voluntary solutions (eg: a code of conduct). Discussions to come to a code of conduct have been initiated at many different levels: in individual food companies, by the industry associations, or by Nutrition Foundations.

A code of conduct, preferably signed by food industries and consumer associations, should contain the following elements:

-a framework which allows monitoring, and actions when non-compliance occurs
-a description of the required scientific underpinning of a health claim

An attempt was made to establish a code of conduct, by the Dutch Nutrition Foundation, based on preliminary work done by a number of important European food industries. In this proposal, the "Code of Conduct for Health Claims" is brought under the umbrella of the General Foods Advertising Code. This code is signed by the food industry, and the consumer association in the Netherlands. The courts may then take the content of the Code to a judge, on complaints.

In this proposal, the scientific basis for a health claim should take into account the following four aspects:

-there should be data from a clinical trial on the product

-the core result of such a clinical trial should have been confirmed at least once

-the results should be peer reviewed, both for their scientific content, as well as for the intended use

-the scientific data should be publicly available at the moment of marketing the product

These four points are found in other, previous, initiatives (eg: one from Sweden, indicating that there is a common basis for discussion among the parties having an interest in functional foods). It is expected that some sort of European solution will be reached in the coming years.

MARINE AND PROCESSED FOODS

Marine Foods

Chapter 5

Functional Seafood Products

Fereidoon Shahidi

Department of Biochemistry, Memorial University of Newfoundland, St. John's, Newfoundland A1B 3X9, Canada

Recognition of the beneficial health effects of marine lipids has resulted in increased consumption of seafoods as such, or their lipids as capsules. Marine lipids are obtained as the liver oil from lean fish, body oil from fatty fish, and blubber oil from marine mammals. Marine oils are a rich source of omega-3 fatty acids with potential nutritional and health benefits in the prevention and possible treatment of cardiovascular diseases (CVD), arthritis, autoimmune diseases and cancer. Meanwhile, marine proteins possess a well-balanced amino acid composition and could be used as such, or as thermostable dispersions and hydrolyzate preparations. The latter products may be used for geriatric foods, athletic drinks and as food components for immune-compromised patients. Furthermore, chitin/chitosan and related compounds produced from shellfish processing discards may be used in a variety of applications. A cursory account of seafood lipids, proteins and chitinous material and their technological, nutritional and beneficial health effect is presented.

The global catch of fish and other aquatic species is approximately 100 million metric tons annually and this has increased in recent years due to production of seafoods through aquaculture. Although seafood proteins are important in that they possess a well-balanced amino acid composition, many species of fish remain underutilized because of a multitude of drawbacks, most of which relate to the small size, high bone content, unappealing shape and look or fatty nature of the species. Therefore, production of novel protein preparations from such raw material has been attempted. These include protein concentrates/dispersions and hydrolyzates, among others. However, seafood lipids which are rich in long-chain omega-3 polyunsaturated fatty acids (PUFA) have attracted much interest and are the main focus of attention. The interest in seafood lipids and their fatty acid constituents stemmed from the observation of the diet of Greenland Eskimos (1) in which fish as well as seal meat and blubber was important. The incidence of cardiovascular disease (CVD) in Eskimos was considerably less than that of the Dane population, despite their high dietary fat intake.

The beneficial health effects of omega-3 PUFA have been attributed to their ability to lower serum triacylglycerol and cholesterol. In addition, omega-3 fatty acids are essential for normal growth and development and may also play a role in the prevention and treatment of hypertension, arthritis, other inflammatory and autoimmune disorders, and cancer (2,3).

It is not only the protein and lipids from aquatic species that have beneficial health effects; chitinous material obtained for shellfish processing discards may be processed into different products for food, pharmaceutical and non-food applications. The production of chitinous materials could potentially reach a minimum of 200,000 metric tons annually. Thus, value-added utilization of exoskeleton of shrimp, crab and lobster is also warranted.

This contribution provides a complementary account of the novel uses of marine proteins and lipids in different applications. Issues related to the stability and stabilization of marine lipids, production of omega-3 concentrates and formulation of other value-added products from underutilized fish species (4) are also discussed. Finally, a cursory account of the production and uses of chitinous material is presented.

Seafood Proteins

Seafood proteins possess excellent amino acid scores and digestibility characteristics. These constitute approximately 11-27% of seafoods and, like those of all other muscle foods, might be classified as sarcoplasmic, myofibrillar and stroma-type. In addition, non-protein nitrogen (NPN) compounds are present, to different extents, depending on the species under consideration. The dark muscles of fish generally contain a higher amount of NPN compounds than their light counterparts.

The sarcoplasmic proteins, mainly albumins, account for approximately 30% of the total muscle proteins; myoglobin and enzymes are also present. The content of sacroplasmic proteins is generally higher in pelagic fish species as compared with demersal fish. Dark muscles of some species contain less sarcoplasmic proteins than their white muscles (5). However, presence of large amounts of myoglobin, hemoglobin and cytochrome C in dark muscles may reverse this trend. Meanwhile, presence of sacroplasmic enzymes is responsible for quality deterioration of fish after death. These include glycolytic and hydrolytic enzymes and their activity depends on the species of fish, type of muscle tissue as well as seasonal and environmental factors.

The myofibrillar proteins in the muscles (approximately 60-70%) are composed of myosin, actin, tropomyosin and troponins C, I and T (5). Myofibrillar proteins undergo changes during the rigor mortis, resolution of rigor and long-term frozen storage. The texture of fish products and the gel-forming ability of fish mince and surimi may also be affected by these changes. The myofibrillar fraction of muscle proteins dictates the general functional properties of muscle tissues.

The residue after extraction of sacroplasmic and myofibrillar proteins is known as stroma which is composed of collagen and elastin from the connective tissues. In general fish muscles contain 0.2-2.2% collagen (6). Although a higher content of collagen contributes to the toughness of the muscle, no such problems are encountered in fish. However, some species of squid may develop a tough and rubbery texture upon heat processing.

The non-protein nitrogen compounds in muscle tissues are composed of free amino acids, amines, amine oxides, guanidines, nucleotides and their breakdown products, urea and quarternary ammonium salts (7). The contribution of NPN compounds to the taste of seafoods is of paramount importance.

Novel Functional Protein Preparations from Underutilized Aquatic Species

Novel functional protein preparations that might be produced from underutilized aquatic species include surimi, low viscosity thermostable proteins and protein hydrolyzates, among others, as discussed below.

Surimi. Surimi is mechanically deboned fish flesh that has been washed with water or very dilute salt solutions at 5-10°C and to which cryoprotectants have been added (8). Removal of water-soluble sarcoplasmic proteins, including hemoproteins, enzymes and NPN compounds results in production of a reasonably colorless and tasteless product. The yield of the process is 50-60% (see Figure 1). Surimi so produced may be used for production of crab-leg analogues and other value-added food products. Colorants and flavorants may be added to surimi in order to form such products.

Factors which affect the surimi quality (gel-forming ability) and yield include storage technique, raw material quality and various processing factors. The process of surimi production from lean (8) and fatty (9) fish species has recently been reviewed and therefore will not be discussed here.

Low Viscosity Thermostable Protein Dispersions. Lack of solubility of myofibrillar proteins in water and their sensitivity to denaturation are among reasons for underutilization of proteins from low-cost fish. Under physiological conditions and in the presence of enzymes and low molecular weight components, myosin molecules interact with one another as well as these compounds (5,10). Thus, developing novel processes for production of dispersion and soluble protein preparations are necessary.

The general process for production of protein dispersions from several species of fish, namely capelin (*Mallotus villosus*), mackerel (*Scomber scombrus*), herring (*Clupea harengus*) and shark (*Isurus oxyrinchus*) is depicted in Figure 2. For capelin, mechanically deboned meat and for other fish, light muscles were comminuted and used as raw material. The preparation of thermostable water dispersions of fish structural proteins, which may also be concentrated by suitable dehydration techniques, involved washing of the comminuted muscle tissues to remove soluble components and lipids. Conformational changes in proteins enhances the entrapment of water in the gel matrices (11). After a preliminary washing of the sample with water, solids were further washed with dilute saline and sodium bicarbonate solution followed by a subsequent wash with cold water. After removal of the water by straining, washed meat was suspended in cold water, at up to 50% and homogenized. In some cases it was necessary to lower the pH to 3.0-3.5 in order to enhance gelation (12). The pH-adjusted dispersion samples were then heated to 50°C. The acetic acid used for this purpose may also act as a preservative during storage of the product prior to its dehydration/concentration.

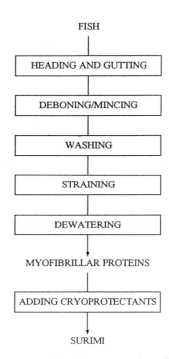

Figure 1. Flowsheet for preparation of surimi.

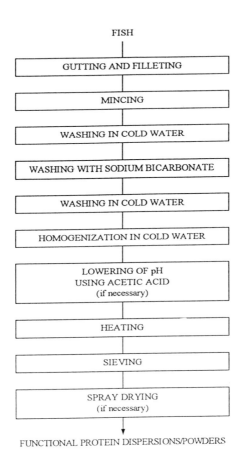

FISH

GUTTING AND FILLETING

MINCING

WASHING IN COLD WATER

WASHING WITH SODIUM BICARBONATE

WASHING IN COLD WATER

HOMOGENIZATION IN COLD WATER

LOWERING OF pH
USING ACETIC ACID
(if necessary)

HEATING

SIEVING

SPRAY DRYING
(if necessary)

FUNCTIONAL PROTEIN DISPERSIONS/POWDERS

Figure 2. Flowsheet for preparation of functional protein dispersions from fish.

Table 1 summarizes the proximate composition of unwashed and washed fish meat before the final homogenization step in the production of thermostable protein products from capelin, herring, mackerel and shark (4,13-15). In all cases examined, the washed meat had a higher moisture and a lower lipid content and had a light color

TABLE 1

Proximate composition (weight %) of unwashed and washed fish meat.

Species	Component	Unwashed	Washed
CAPELIN	Moisture	83.98 ± 0.20	93.19 ± 0.10
	Crude Protein (N x 6.25)	12.70 ± 0.31	4.51 ± 0.15
	Lipid	1.98 ± 0.01	1.13 ± 0.04
HERRING	Moisture	72.4 ± 0.10	86.7 ± 2.30
	Crude Protein (N x 6.25)	17.4 ± 1.20	8.30 ± 0.20
	Lipid	8.10 ± 1.20	4.10 ± 0.10
MACKEREL	Moisture	63.6 ± 0.30	82.20 ± 1.60
	Crude Protein (N x 6.25)	19.30 ± 0.20	9.60 ± 1.10
	Lipid	4.0 ± 0.50	5.30 ± 0.20
SHARK	Moisture	79.8 ± 0.11	91.54 ± 0.16
	Crude Protein (N x 6.25)	18.25 ± 0.32	7.58 ± 0.51
	Lipid	1.03 ± 0.05	0.55 ± 0.01

and a bland taste. The treatment also increased the mass of the sample by about 20% (w/w), presumably due to increased hydration of proteins.

The aqueous dispersions of washed meats from herring, mackerel and shark were highly viscous and their viscosity was both temperature and concentration dependent. As the temperature of the dispersions increased, their viscosity decreased, however, upon subsequent cooling the apparent viscosity of the dispersion was mostly recovered. In addition, the viscosity of the herring and mackerel dispersions was increased to > 5 Pa·s when 2.4% proteins were present. For capelin dispersions, heating of the samples to 80-100°C resulted in a permanent loss of viscosity. However, for herring, mackerel, and shark dispersions loss of viscosity was achieved only when a small amount of acetic acid was added to the washed samples prior to heat treatment. Thus, the solubility of protein dispersions, once formed, was ≥ 83% in all cases examined (Results not shown).

The dispersions with reduced apparent viscosity showed remarkable thermal stability. The proteins were stable when heated to 100°C for 30 min and even moderate concentrations of salt in the solution did not influence the solubility characteristics of such dispersions. Stable proteins are of interest to study structure-function relationships and to examine possibilities of their use in biological applications (16). Recently, Wu et al. (17) and Stanley et al. (18) have reported solubility of beef and chicken myofibrillar proteins in low ionic strength media and Doi (19) demonstrated that under controlled pH, ionic strength and heating, one could obtain transparent gels from globular protein which may have potential application in food preparations. Use of these dispersions as a protein supplement in extrusion cooking of cereal-based products and development of a functional powder by suitable dehydration of the dispersion is feasible.

The fish protein concentrates (powders) prepared from different species in this study, following dehydration, had a protein content of ≥ 85%. The amino acid composition of products so obtained was similar to that of the proteins present in the starting muscle tissues. However, the content of individual free amino acids in the products was by an order of magnitude lower than those present in the starting materials. Thus, the dispersions and powders so prepared had the advantage of being of equal nutritional value to their original fish proteins but having a bland taste. In addition, the washing process removed most of the hemoproteins from the original muscle tissues and products so obtained had a milky white colour. Use of such dispersions for fortification of cereal-based products is promising (20).

Protein hydrolyzates. Freshly landed aquatic species such as male capelin (*Mallotus villosus*) may be mechanically deboned, similar to the procedure employed in surimi production. The comminuted raw materials is then suspended in water and enzyme is added to the slurry (21). The reaction may be allowed to proceed between 2h and 1 week, depending on the activity of the enzyme employed as well as the process temperature and other factors. After separation of solids, the aqueous layer is clarified, pH adjusted and the material dehydrated (Figure 3). The process may include sterilization at different stages, if necessary.

Typical protein yield and proximate composition of hydrolyzates from capelin (Biocapelin) and shark (Bioshark) are given in Table 2. Although many factors affect the yield of hydrolysis, the type of enzyme employed had a marked effect on this and also the characteristics of the final product. The enzymes examined in the present investigation included papain, Alcalase, Neutrase and endogenous autolytic enzymes of each species. High yields of protein recovery by Alcalase and its low cost may provide an incentive for its use in commercial operations (22).

A close scrutiny of the results presented in Table 2 indicates that hydrolyzates generally possess a lower lipid content than their original protein source, on dry weight basis. As hydrolysis proceeds, elaborate membrane system of the muscle cells tend to round up and form insoluble vesicles, thus allowing the removal of membrane structural lipids. Consequently, protein hydrolyzates are expected to be more stable towards oxidative deterioration. In addition, hydrolyzates produced from capelin and seal meat had an ivory-white colour which might have significance when formulating crab leg analogues and other fabricated products to which addition of colorants is generally desirable.

The amino acid composition of capelin protein hydrolyzate (Biocapelin) and shark protein hydrolyzate (Bioshark) produced by Alcalase-assisted hydrolysis was compared with those of their corresponding muscle tissues. Results indicated that the amino acid profiles for both species remained generally unchanged. However, sensitive amino acids such as methionine and tryptophan were slightly affected. Furthermore, the content of free amino acids in the hydrolysates was increased from 123 mg/100 g sample to 2,944 mg/100 g. Thus, the hydrolysates so prepared may be used for thermal generation of aroma when combined with sugars such as glucose (23).

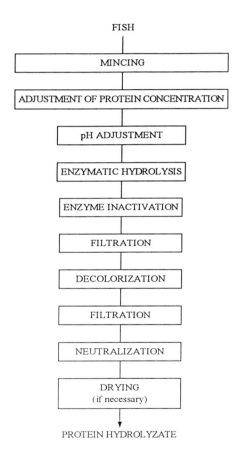

Figure 3. Flowsheet for production of protein hydrolyzates.

The hydrolyzates prepared from capelin and shark had excellent solubility characteristics at pH values ranging from 2.0 to 10.4. While 90.4-98.6% of nitrogenous compounds of Biocapelin were soluble, corresponding values for Bioshark varied between 95.5 and 98.1%. The fat adsorption, moisture retention, emulsification properties and whippability of the hydrolyzates were also excellent (15,24). In addition, when capelin and shark protein hydrolyzates were added to meat model systems at 0.5-3.0% levels, an increase of up to 4.0% and 8.0%, respectively, in the cooking

TABLE 2

Typical protein recovery (%) and composition of hydrolyzates from capelin (Biocapelin) and shark (Bioshark).

Material	Protein Recovery	Composition, %			
		Protein	Lipid	Moisture	Ash
Capelin	–	13.6-14.1	3.3-3.9	78.0-78.3	2.4-2.5
Biocapelin	51.6-70.6	65.9-73.3	0.2-0.4	5.3-6.3	14.9-20.6
Shark	–	17.9-18.6	1.0-1.1	79.6-80.0	0.8-0.9
Bioshark	58.0-72.4	77.4-78.0	0.2-0.3	7.4-8.2	13.4-14.3

yield of processed meats was noticed (15,24). Furthermore, Biocapelin inhibited oxidation of meat lipids by 17.7-60.4% as reflected in the content of 2-thiobarbituric acid reactive substances of the treated meat systems. The mechanism by which this antioxidant effect is exerted may be related to the action of hydrolyzed molecules as chelators of metal ions that could be released from hemoproteins during the heat processing of the meat. Furthermore, the phenolic nature of some amino acids such as tyrosine may allow their participation as free radical scavengers in the medium (25,26).

Marine Lipids and their Beneficial Health Effects

Marine lipids originate from the liver of lean white fish such as cod, the body of oily fish such as mackerel, and the blubber of marine mammals such as seal. These oils consist of saturated, monounsaturated and polyunsaturated fatty acids (PUFA). There are two classes of PUFA, namely the omega-3 and the omega-6 families and these are differentiated from one another based on the location of the double bond from the terminal methyl group of the fatty acid molecule. Unlike saturated and monounsaturated fatty acids which can be synthesized by all mammals, including humans, the PUFA cannot be easily synthesized in the body and must be provided through the diet. The omega-3 family of PUFA are descended from linolenic acid while their omega-6 counterparts are descended for linoleic acid. The unique feature that differentiates lipids of marine species from land animals is the presence of long-chain PUFA, namely eicosapentaenoic acid (EPA; C20:5 ω3), docosahexaenoic acid (DHA; C22:6ω3) and, to a lesser extent, docosapentaenoic acid (DPA; C22:5 ω3). These PUFA are formed in unicellular phytoplankton and multicellular sea algae and eventually pass through the food web and become incorporated into the body of fish and other higher marine species (27). The high content of omega-3 fatty acids in marine lipids is suggested to be a consequence of cold temperature adaptation in which omega-3 PUFA remain liquid and oppose any tendency to crystallize (28).

Illingworth and Ullman (29) proposed that consumption of marine oils results in a decrease in plasma lipids by reduced synthesis of fatty acids and very low density lipoproteins (VLDL). Nestel (30) suggested that the long-chain omega-3 PUFA have a direct effect on the heart muscle itself, increase blood flow, decrease arrhythmias, improve arterial compliance, decrease the size of the infarct and reduce several cellular processes that comprise heart function. It has also been suggested that marine oils may retard artherogenesis through their effects on platelet function, platelete-endothelial interactions and inflammatory response. Most of these effects are mediated, at least in part, by alterations in eicosenoids formation in the human body (31).

The parents of the two PUFA families, namely linoleic and linolenic acids (see above) are the precursors of long chain PUFA which in turn produce a range of regulatory and biologically active substances, collectively known as eicosanoids. In general, compounds derived from the omega-3 PUFA are less powerful in their effects than those derived from the omega-6 PUFA. Both the omega-3 and omega-6 pathways operate through the same set of enzymes and may compete with each other. Consumption of high amounts of dietary marine oils leads to the production of the less powerful compounds.

Both arachidonic acid (AA) and EPA are precursors of eicosanoids (32); such as prostaglandins, thromboxanes and leukotriens, all of which are oxygenated derivatives of C20 fatty acids. Eicosanoids from these two fatty acids are different in structure and function (31) and have a broad spectrum of biological activity. Physiological effects of omega-3 fatty acids have been observed in the areas of i) heart and circulatory, ii) immune response and iii) cancer. The first group includes prevention or treatment of atherosclerosis (33,34), thrombosis (2), hypertriacylglycidemia (35), and high blood pressure (33). The second area relates to the treatment of asthma, arthritis (36), migraine headache, psoriasis and nephritis (2). The third category involves cancer of the breast (32), prostate and colon (36).

Since eicosanoids are ultimately derived from PUFA in the diet, it is clear that both qualitative and quantitative changes in the supply of dietary PUFA will have a profound effect on the production of eicosanoids. Therefore, omega-3 fatty acids are emerging as being essential nutrients. As a structural component of brain, retina, testes and sperm, DHA appears to be linked to proper tissue function and needs to be supplied in sufficient amounts during tissue development (37). Recent studies have shown that DHA supplementation during pregnancy and lactation is necessary in order to prevent deficiency of the mother's DHA status during the periods in order to meet the high fetal requirement for DHA (38). Carlson et al. (39) have shown that premature babies have lower levels of DHA in their tissues as compared to those of full-term babies. Thus, supplementation of infant formula with marine oils/DHA is necessary in order to provide them with as much DHA as that available to their breast-fed counterparts. Feeding of infants with formula devoid of omega-3 fatty acids resulted in lack of deposition of DHA in their visual and neural tissues (40).

The Canadian Scientific Review Committee of Nutrition Recommendations (41) has suggested daily requirements for PUFA (omega-3 and omega-6) based on energy needs. Additional amounts of omega-3 and omega-6 fatty acids are recommended for lactating and pregnant women with increasing amounts from first to the second trimester of pregnancy. British Nutrition Foundation Task Force (42) on unsaturated fatty acids has recommended that 5% of total daily energy supply for humans should originate from omega-3 fatty acids.

Processing of Marine Oils
The basic processing steps for production of marine oils depends on the source of raw material used, but generally includes heat processing or rendering to release the oil, degumming, alkali-refining, bleaching and deodorization. Thus, processing of marine oils is similar to that of vegetable oils (43). However, the quality of crude marine oils is less uniform than crude vegetable oils and proper handling of the raw material is of utmost importance. In addition, marine oils must be processed at the lowest possible temperature and under a blanket of nitrogen, when possible. Figure 4 provides a flow diagram to exhibit different processing steps involved in production of marine oils.

Oxidation of Marine Oils
Oils rich in PUFA are highly prone to oxidative deterioration and easily produce off-flavors and off-odors (24,44). Therefore, inhibition of oxidation is a major criterion when marine oils are processed, stored or incorporated into food products.

Oxidation of marine oils proceeds via a free-radical chain mechanism involving initiation, propagation and termination steps. Peroxides are the primary products of oxidation, but these toxic compounds are unstable and degrade further to secondary products such as malonaldehyde and 4-hydroxynonenal that are also highly toxic. Secondary oxidation products are responsible for the development of off-flavor in stored seafoods and marine oils. Therefore, every attempt should be made to control deteriorative processes of oxidation in order to obtain maximum benefit from their omega-3 components.

Methods that might be used to stabilize marine lipids include addition of antioxidants, microencapsulation and hydrogenation. However, hydrogenation negates the beneficial health effects of PUFA by producing saturated fatty acids as well as trans-isomers. Therefore use of antioxidations and microencapsulation technology might be practiced.

Antioxidants. Synthetic antioxidants such as butylated hydroxyanisole (BHA), butylated hydroxytoluene (BHT), propyl gallate (PG) and tert-butylhydroquinone (TBHQ) as well as synthetic α-tocopherol may be incorporated into seafoods and/or marine oils in order to control their oxidative deterioration. However, the effectiveness of synthetic antioxidants other than TBHQ in prevention of oxidation of marine oils is very limited. In addition, synthetic antioxidants are regarded as suspects in cancer development when used at high concentrations in experimental animals. As a result, BHA has been removed from the GRAS (generally recognized as safe) list by the US Food and Drug Administration (FDA; 45). Furthermore, TBHQ has not been approved for use in Europe, Japan and Canada. Therefore, natural antioxidants may offer a practical solution for control of oxidation of marine oils. Although antioxidants have traditionally been used to stabilize lipids, they may also augment body antioxidant defence system for combating cancer and degenerative diseases of ageing. The natural antioxidants from dietary sources generally belong to the phenolic group of compounds, including tocopherols and flavonoids, as well as phospholipids and polyfunctional

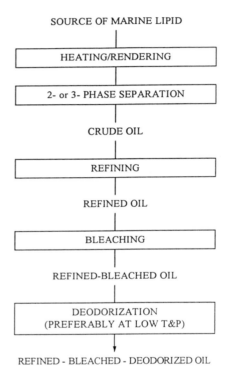

Figure 4. Process for production of crude, refined, refined-bleached, and refined-bleached-deordized marine oils.

organic acids. Use of rosemary extract, plant tocopherols, oat oil, lecithin and ascorbates, individually or in specific mixtures and combinations may be encouraged in order to arrest oxidation.

Among the natural sources of antioxidants, green tea extracts might provide a practical and highly effective means for controlling oxidation and may also have the advantage of exerting beneficial health effects in prevention of cancer. Thus, inclusion of green tea extracts in marine oils may afford products which are effective in the control and possible treatment of both CVD and cancer.

Recent studies in our laboratories have indicated that while ground green tea leaves and crude extracts from green tea are effective in inhibiting oxidation of cooked ground light muscles of mackerel (see Table 3), they may require complete dechlorophillization to be useful for incorporation into marine oils (46). Meanwhile, the activity of individual catechins tested was in the order of epicatechin gallate (ECG) $\approx$ epigallocatechin gallate (EGCG) > epigallocatechin (EGC) » epicatechin (EC). However, ground green tea leaves protected the cooked fish meat against oxidation more effectively than individual catechins, perhaps due to the presence of other active components which might have acted synergistically with catechins. Further studies on the use of dechloriphillized green tea extracts (DGTE) and individual catechins revealed their efficacy in inhibiting oxidation of marine oils. These results, as reflected in the TBARS values of stored menhaden and seal blubber oils under Schaal oven conditions, are summarized in Table 4.

TABLE 3

The content 2-thiobarbituric acid reactive substances (mg malonaldehyde equivalents/kg sample) of cooked ground light muscles of mackerel treated with different antioxidants over a 7-day storage period at 4°C.[a]

Antioxidant (ppm)	Fresh	Stored
None	8.81	12.03
α-Tocopherol (296)	7.34	9.76
GGTL (1265)	4.09	4.25
GTE (1265)	4.03	5.58
EC (200)	5.81	6.53
EGC (211)	3.91	4.46
ECG (304)	3.38	3.90
EGCG (316)	3.29	3.84

[a]Based on mole equivalents of 200 ppm epicatechin, EC. Other symbols are: GGTL, ground green tea leaves; GTE, green tea extract; EGC, epigallocatechin, ECG, epicatechin gallate; and EGCG, epigallocatechin gallate.

Microencapsulation. Oils rich in polyunsaturated fatty acids may be stabilized by their inclusion in microcapsules. In this process, the oil is entrapped in a wall material and is protected for oxygen, moisture and light (47). The initial step in encapsulation of a food ingredient is the selection of a suitable coating matter. Wall materials are basically

42

film-forming substances which can be selected from a wide variety of natural or synthetic polymers. This depends on the nature of the material to be coated and the desired characteristics of the final microcapsules. Therefore, carbohydrates, both as such or modified, as well as proteins may be used for this purpose. As an example, Taguchi et al. (48) have reported microencapsulation of sardine oil in egg white powder while Lin et al. (49) used gelatin and caseinate for microencapsulation of squid oil.

TABLE 4
Percent mean inhibition of formation of 2-thiobarbituric acid reactive substances in marine oils as affected by antioxidants.[1]

Antioxidant (ppm)	Menhaden oil	Seal blubber oil
α-Tocopherol (200)	14.14	13.24
BHA (200)	25.28	22.97
BHT (200)	31.37	35.47
TBHQ (200)	51.80	56.25
DGTE (100)	25.25	22.97
DGTE (200)	32.91	32.91
DGTE (500)	41.44	41.81
DGTE (1000)	44.64	47.11
EC (200)	41.40	39.53
EGC (200)	46.46	40.50
ECG (200)	52.85	58.59
EGCG (200)	49.41	50.03

[1]Symbols are: BHA, butylated hydroxyanisole; BHT, butylated hydroxytoluene; TBHQ, tert-butyl hydroquinone; DGTE, dechlorophillized green tea extract, EC, epicatechin; EGC, epigallocatechin; ECG, epicatechin gallate; and EGCG, epigallocatechin gallate.

In a recent study (50) we used β-cyclodextrin, Maltrin and corn syrup solids for microencapsulation of seal blubber oil. The efficacy of well materials to prevent oxidation of the resultant coated products was in the order of β-cyclodextrin >> corn syrup solids > Maltodextrin. It is possible that inclusion of alkyl chains of PUFA in the central cavity of β-cyclodextrin provides better protection to the oil (Table 5).

Omega-3 Concentrates
Omega-3 concentrates from marine oils may be produced in the form of the natural triaylglycerols or as modified triacylglycerols, as free fatty acids, or as the simple alkyl esters of these acids. Winterization or fractional crystallization may be employed to prepare concentrates with a total EPA, DHA and DPA content of up to 30%. However, production of higher total concentration of the fatty acids is difficult because of the complexity of the combination of the fatty acids in triacylglycerol oils. However, these omega-3 concentrates might be prepared easily once they are transformed to free fatty acids or their simple alkyl esters. Urea complexation is one of the most promising methods for concentrating PUFA or their esters as it allows handling of large quantities

of material in simple equipment (51). In this process saturated and longer chain monounsaturated fatty acids are complexed with urea as inclusion compounds while PUFA remain in the mixture and can be separated from it. In addition, lipase assisted hydrolysis, alcoholysis and acidolysis may allow selective concentration of omega-3 fatty acids. The enzymes used for production of omega-3 concentrates may show selectivity for the position of acyl groups in the triacylglycerol molecule and also the degree of unsaturation of these acids. The degree and yield of concentration depends on the positional distribution of fatty acids in the triacylglycerol molecules.

TABLE 5
Changes in the content (%) of polyunsaturated fatty acids of encapsulated
(in β-cyclodextrin) seal blubber oil upon storage.[a]

Fatty acid	Initial	3-Weeks	7-Weeks
EPA	7.5	7.1 (3.3)	7.0 (3.0)
DPA	4.9	4.2 (2.1)	4.2 (2.1)
DHA	8.4	8.0 (3.8)	7.2 (3.0)
Total Polyunsaturates	24.7	23.3 (13.1)	22.2 (11.5)

[a] Symbols are: EPA, epicosapentaenoic acid; DPA, docosapentaenoic acid; and DHA, docosahexaenoic acid. Values in parentheses are for unencapsulated oils. Adapted from reference 50.

TABLE 6
Omega-3 fatty acids of seal blubber oil concentrates.[a]

Fatty Acid	Original Oil	Concentrate	
		Enzymatic	Urea Complexation
EPA	9.1	12.2 - 17.1	24.6-27.7
DPA	5.0	6.9 - 8.4	5.0-9.3
DHA	10.0	15.4 - 26.6	38.9-46.1
Total Polyunsaturates	24.7	54.1	88.2

[a] Symbols are: EPA, eicosapentaenoic acid; DPA, docosapentaenoic acid; and DHA, docosahexaenoic acid.

Table 6 shows the degree of concentration of omega-3 fatty acids from seal blubber oil using enzyme-assisted hydrolysis and urea complexation procedures. The total content of omega-3 fatty acids from the enzymatic process, under optimum conditions, was approximately 54% and that for urea complexation process was 88%.

Chitin-based products
The seafood processing industry produces large amounts of shell waste which could be used for production of chitin as well as other value-added products. Chitin is a natural biopolymer which is similar to cellulose, but in which the 2-hydroxy group in the glucose

monomer is replaced with a 2-acetamino group. If deacetylated, chitosan is produced which has wide-spread applications in different industries. The content of chitin in shell wastes accounts for up to 50% of the raw material on a dry weight basis. Figure 5 provide a flowsheet for production of chitin, chitosan and related products from exoskeleton of crab, shrimp and lobster (52,54).

Chitin is insoluble in almost every common organic solvent and in acidic, basic and neutral aqueous solutions. However, chitosan is insoluble in dilute acids and in aqueous solutions. Production of chitosan follows unit operations of deacetylation, washing and drying (see Figure 5). The quality characteristics of chitosans depend, to a large extent, on the degree of deacetylation of the macromolecule. The crude chitosan may be further purified by dissolution in a suitable acid solution followed by filtration and precipitation by pH adjustment. Chitosan derivatives with different functionalities may be prepared in order to address specific needs of the under industries.

Table 7 summarizes some of the numerous applications of chitinous materials in different areas of food, agriculture, biomedical, biotechnology, and water treatment, among others (55-60). However, such applications may face a serious challenge due to competitions from synthetic material.

A novel approach for utilization of chitinous materials was recently examined in our laboratories. N,O-Carboxymethylchitosan (NOCC) was prepared in a laboratory scale or obtained from Novo Chem. Inc. (Halifax, NS). It was noted that NOCC and its lactate, acetate and pyrrolidine carboxylate salts were able to prevent cooked meat flavour deterioration over a nine-day storage period at refrigerated temperatures. The mean inhibitory effect of NOCC and its aformational derivatives at 0.05-0.3% addition level in the formation of oxidative products as reflected in 2-thiobarbituric acid reactive substances was 46.7, 69.9, 43.4 and 66.3%, respectively. The mechanism by which this inhibition takes place is thought to be related to chelation of free iron which is released from hemoproteins of meat during heat processing. This would in turn inhibit the catalytic activity of iron ions.

Depolymerization of chitin produces N-acetylglucosamine (NAG). It has been reported that NAG has anti-inflammatory effect and may possess interesting characteristics. Although chemical preparation of NAG is feasible, a commercially attractive process for its production would be much desirable. Therefore, consideration of reaction conditions under high pressure may prove beneficial. Use of NAG in different synthetic applications may also be of interest to biochemists. In addition, the sulfate salt of decacelylated form of NAG, namely glucosamine sulphate may be used for treatment of arthritis and its effect may be exentuated by simultaneous oral administration of marine oils. Therefore, consumption of marine-derived products not only provides nutritional products for humans, but also have beneficial effects in control of many chronic diseases.

Conclusions

Novel protein, and chitinous lipid products may be prepared from underutilized aquatic species and their processing discards. These products may be used in pharmaceutical, food and non-food specialty applications. The beneficial health effects of seafood lipids, chitinous materials and the nature of seafood proteins would provide the necessary incentive for consumers to modify their current dietary habits, particularly in the Western world.

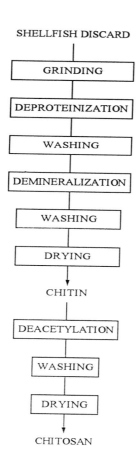

SHELLFISH DISCARD

GRINDING

DEPROTEINIZATION

WASHING

DEMINERALIZATION

WASHING

DRYING

CHITIN

DEACETYLATION

WASHING

DRYING

CHITOSAN

Figure 5. A simplified flowsheet for production of chitin and chitosan.

Table 7. Some application of chitin and its derivatives.

Area of Application	Examples
Food	- Clarification of wine and juice - Dietary fibre and edible films - Protein flocculation - Removal of tannins - Chromatography
Agriculture	- Coating for delayed ripening of fruits - Seed coating - Nutrient control release - Nematode treating - Animal feed
Water Treatment	- Food processing - Potable drinking water - Removal of dyes - Removal of metals, pesticides and PCBs - Sewage treatment
Additive	- Inhibition of oxidation - Thickener - Stabilizer/emulsifier - Texture modifier - Slow-release additive support
Biomedical	- Hypocholesterolemic effect - Wound care - Eye bandage - Drug delivery - Biomaterials - Dental applications
Biotechnology	- Enzyme immobilization - Cell immobilization - Encapsulation - Filter aid - Protein recovery

Literature Cited

1. Bang, H.O.; Dyerberg, J.; Hjorne, N. **1976**, *Acta Med. Scand. 200*, 69-73.
2. Kinsella, J.E. **1986**, *Food Technol. 40 (2)*, 89-97.
3. Simopoulos, A.P. **1981**, *Am. J. Clin. Nutr. 54*, 438-463.
4. Venugopal, V.; Shahidi, F. **1995**, *CRC Crit. Rev. Food Sci. Nutr. 35*, 431-435.
5. Suzuki, T. **1981**, Fish and Krill Processing Technology, Applied Science Publishers, London.

6. Sato, K.; Yoshinaka, R.; Sato, M.; Shimizu, Y. **1986**, *Bull. Jpn. Soc. Sci. Fish.* *52*, 1595-1598.

7. Ikeda, S. **1979**, In *Advances in Fish Science and Technology*, Connell, J.J. (Ed.), Fishing News Books, Surrey, pp. 111-124.

8. Lee, C.M. **1994**, Surimi processing from lean fish. In *Seafoods: Chemistry, Processing Technology and Quality.* Shahidi, F. and Botta, J.R. (Eds.), Blackie Academic and Professional, Glasgow. pp. 263-287.

9. Spencer, K.E.; Tung, M.A. **1994**, In *Seafoods: Chemistry, Processing Technology and Quality*, Shahidi F. and Botta, J.R. (Eds.). Blackie Academic and Professionals, Glasgo, pp. 288-319.

10. Nakagawa, T.; Nagayama, F.; Ozaki, H.; Watabe, S.; Hashimoto, K. **1989**, Effect of glycotytic enzymes on the gel forming ability of fish muscle. Nippon Suisan Gakkaishi *55*, 1945.

11. Ziegler, G.R.; Foegeding, E.A. **1990**, In *Advances in Food Research*, Volume 34, J.E. Kinsella (ed.), Academic Press, New York, pp. 203-298.

12. Freitheim, K.; Egelandsdal, B.; Harbitz, O.; Samejima, K. **1985**, *Food Chem.* *18*, 169-177.

13. Shahidi, F.; Venugopal, V. **1993**, *Meat Focus International 2*, 443-445.

14. Shahidi, F.; Onodenalore, A.C. **1995**, *Food Chem. 53*, 51-54

15. Onodenalore, A.C.; Shahidi, F. **1997**, *J. Aquatic Food Prod. Technol.* In press.

16. Nosoh, Y.; Sekiguchi, T. **1991**, In *Protein Stability and Stabilization through Protein Engineering.* Ellis Horwood Ltd., Sussex, England, pp. 101-123.

17. Wu, Y.J.; Atallah, M.T.; Hultin, H.O. **1992**, *J. Food Biochem. 15*, 209-218.

18. Stanley, D.W.; Stone, A.P.; Hultin, H.O. **1994**, *J. Agric. Food Chem. 42*, 863.

19. Doi, E. **1993**, *Trends Food Sci. Technol. 4*, 1.

20. Venugopal, V.; Doke, S.N.; Nair, P.M.; Shahidi, F. **1994**, *Meat Focus International. 3*, 200-202.

21. Mohr, V. **1977**, In *Biochemical Aspects of New Protein Food*, Adler-Nissen, J., Eggum, B.O., Munck, L. and Olsen, H.S. (eds.), FEBS Federation of European Biochemical Societies, 11th Meeting, Copenhagen, Vol. *44*, pp. 259-269.

22. Shahidi, F.; Han, X.-Q.; Synowiecki, J. **1995**, *Food Chem. 53*, 285-293.

23. Ho, C.-T.; Chen, C.-W.; Wanasundara, U.N.; Shahidi, F. **1997**, In *Natural Antioxidants: Chemistry, Health Effects and Applications.* Shahidi, F. (Ed.). AOCS Press, Champaign, IL. pp. 213-223.

24. Shahidi, F.; Synowiecki, J.; Balejko, J. **1994**, *J. Agric. Food Chem. 42*, 2634-2638.

25. Shahidi, F.; Amarowicz, R. **1996**, *J. Am. Oil Chem. Soc. 73*, 1197-1199.

26. Amarowicz, R.; Shahidi, F. **1997**, *Food Chem. 58*, 355-359.

27. Yongmanichai, W.; Ward, O.P. **1989**, *Prog. Biochem. 24*, 117-125.

28. Ackman, R.C. **1988**, The year of fish oil. *3*, 139-145.

29. Illingworth, D.; Ullmann, D. **1990**, In *Omega-3 Fatty Acids in Health and Disease.* R.S. Lees and M. Karel (eds.). Marcel Dekker, Inc., New York, NY, pp. 39-69.

48

30. Nestel, P.J. **1990**, N-3 fatty acids, cardiac function and cardiovascular survival. Paper presented at the 2nd International Conference on the Health Effects of Omega-3 Polyunsaturated Fatty Acids in Seafoods. March 20-23, Washington, D.C.

31. Fischer, S. **1989**, *Adv. Lipid Res. 23*, 169-198.

32. Branden, L.M.; Carroll, K.K. **1986**, *Lipids 21*, 285-288.

33. Dyerberg, J. **1986**, *J. Nutr. Rev. 44*, 125-134.

34. Mehta, J.; Lopez, L.M.; Lowton, D.; Wargovich, T. **1988**, *Am. J. Med. 84*, 45-52.

35. Phillipson, B.E.; Rothrock, D.W.; Cunner, W.E.; Harris, W.S.; Illingworth, D.R. **1985**, *New Eng. J. Med. 312*, 1210-1216.

36. Singh, G.; Chandra, R.K. **1988**, *Prog. Food Nutr. Sci. 12*, 371-419.

37. Neuringer, M.; Anderson, G.J.; Conner, W.E. **1988**, *Ann. Rev. Nutr. 8*, 817-821.

38. Al, M.D.M.; Hornstra, G.; Schouw, Y.T.; Bulstra-Remakers, M.E.E.W.; Huistes, G. **1990**, *Early Hum. Dev. 24*, 239-248.

39. Carlson, S.E.; Rhodes, R.G.; Ferguson, M.G. **1986**, Docasahexaenoic acid status of preterm infants at birth and following feeding with human milk or formula. *Am. J. Clin. Nutr. 44*, 798-802.

40. Carlson, S.E.; Rhodes, P.G.; Rao, V.S.; Goldgar, D.E. **1987**, *Pediatr. Res. 21*, 507-511.

41. Canadian Scientific Review Committee, Nutrition Recommendation: Minister of National Health and Welfare, Ottawa, ON, 1990 (H49-42/1990E).

42. British Nutrition Foundation. **1992**. Report of the Task Force on Unsaturated Fatty Acids. Chapman & Hall, London.

43. Bimbo, A.P.; Crowther, J.B. **1991**, *Infofish Int. 6*, 20-25.

44. Ke, P.J.; Ackman, R.G.; Linke, B.A. **1975**, Autoxidation of polyunsaturated fatty compounds in mackerel oil: Formation of 2,4,7-decatrianols. *J. Am. Oil Chem. Soc. 52*, 349-353.

45. Neito, S.; Garrido, A.; Sanhueza, J.; Loyola, L.A.; Morales, G.; Leighten, F.; Velenzuela, A. **1993**, *J. Am. Oil Chem. Soc. 70*, 773-778.

46. Shahidi, F.; Wanasundara, U.N.; He, Y.; Shukla, V.K.S. **1997**, In *Flavor and Lipid Chemistry of Seafood.* ACS Symposium Series. American Chemical Society, Washington D.C., in press.

47. Shahidi, F.; Han, X.-Q. **1993**, *CRC Crit. Rev. Food Sci. Nutr. 33*, 501-547.

48. Taguchi, K.; Iwami, K.; Ibuki, F.; Kawabata, M. **1992**, *Biosci. Biotech. Biochem. 56*, 560-563.

49. Lin, C.-C.; Lin, S.-Y.; Hwang, W.S. **1995**, *J. Food Sci. 60*, 36-39.

50. Wanasundara, U.N.; Shahidi F. **1995**, *J. Food Lipids. 2*, 73-86.

51. Shahidi, F.; Amarowicz, R.A.; Synowiecki, J.; Naczk, M. **1993**, In *Developments in Food Engineering*, Yano T., Matsuno, R. and Nakamura, K. (eds.), Blackie Academic and Professional, London, pp. 627-629.

52. Shahidi, F.; Synowiecki, J. **1992**, In *Advances in Chitin and Chitosan, Brine, C.J., Sanford, P.A. and Zikakis* (Eds.). Elsevier Applied Science, London and New York, pp. 617-626.

53.   Alimunir, A.; Zinuddin, R. **1992**, In *Advances in Chitin and Chitosan, Brine, C.J., Sanford, P.A. and Zikakis* (Eds.). Elsevier Applied Science, London and New York, pp.627-632.

54.   Shahidi, F.; Synowiecki, J. **1991**, *J. Agric. Food Chem. 39*, 1527-1537.

55.   Brzeski, M.J. **1987**, *INFOFISH Inter. 5*, 31-33.

56.   Knorr, D. **1984**, *Food Technol. 38(1)*, 85-97.

57.   Knorr, D. **1991**, *Food Technol. 45(1)*, 114-122.

58.   Michihiro, S.; Watanabe, S.; Kishi, A.; Izume, M.; Ohtakara, A. **1988**, *Lipids. 23*, 187-191.

59.   No, H.K.; Meyers, S.P. **1989**, *J. Agric. Food Chem. 37*, 580-583.

60.   Pandya, Y.; Knorr, D. **1991**. Process Biochem. 26, 25-81.

Chapter 6

Effect of *Crassostera gigas* Extract (JCOE) on Cell Growth in Gastric Carcinoma Cell Lines

Yasuharu Masui, Toshikazu Yoshikawa, Yuji Naito, Yoshio Boku, Takaaki Fujii, Hiroki Manabe, and Motoharu Kondo

First Department of Internal Medicine, Kyoto Prefectural University of Medicine, 465 Kajii-cho, Kawaramachi-Hirokoji, Kamigyo-ku, Kyoto 602, Japan

We evaluated the effect of JCOE (Japan clinic oyster extract), a powder extracted from *Crassostera gigas,* on growth in gastric carcinoma cell lines: MKN-45; a wild type p53, MKN-28; a mutant type p53. Exposure of MKN cells (a cell-line established from human gastric cancer) to JCOE for 72 hours inhibited MKN-45 cell growth in a dose-dependent manner, but not MKN-28 cells. JCOE significantly increased the S peak in the cell cycle of MKN-45 on the DNA histogram in a dose-dependent manner, which indicates that JCOE induces the cell cycle arrest at the peak, of late S or early G2/M. In the Morphological experiments, JCOE did not induce the immediate death of MKN-45 cells by necrosis, but did the protracted death by apoptosis. It may be implicated that the inhibition of MKN-45 cell growth was induced by cell cycle arrest and/or apoptosis but not by necrosis. Neither the cell growth nor the cell cycle of the MKN-28 cells was affected by JCOE. The cause of these differences most likely resulted in the status of the p53 gene of MKN cells, a wild type in MKN-45 cells, and a mutant type in MKN-28 cells.

Chemoprevention, the prevention of cancer by ingestion of chemical agents that reduce the risk of carcinogenesis, is one of the most direct ways to reduce morbidity and mortality from cancer. In searches for new cancer chemopreventive agents over the past several years, hundreds of extracts from natural foods have been evaluated for their potential to inhibit cancer cell growth (1)-(4). Epidemiologists have reported that 80% of human cancers are induced by environmental factors, and that most human cancers can be prevented if the main causes and protective factors against cancers can be identified. Recent evidence has implicated that free radicals play a major role in the initiation and promotion of neoplastic transformation, mutation, and the activation of specific oncogenesis. Therefore, natural foods with free radical scavenging activities are thought to be one of candidates for cancer chemoprevention. JCOE (Japan clinic oyster extract) is an amino-acid powder extracted from *Crassostera gigas*. JCOE was prepared from fresh raw oysters, heated at 80°C for 1 hour, that contains a mixture shown in Table 1. According to the report by the Japan Food Analysis Center, JCOE contains a variety of substances having antioxidant activity, including thiol containing amino acids, especially taurine (5.11mg per 100mg). Our recent *in vitro* study (5) has demonstrated that JCOE directly scavenged not only superoxide radicals, but also hydroxyl radicals in a dose dependent manner. In this study, we investigated the effect of JCOE on cell growth in gastric carcinoma cell lines.

Materials and Methods

Cell growth inhibition   MKN-28 cells derived from moderately differentiated tubular adenocarcinoma, with a mutant p53, and MKN-45 cells derived from poorly differentiated adenocarcinoma, with a wild p53, which cell-lines are established from human gastric cancer, were cultured in RPMI 1640 medium supplemented with 10% FCS (Fatal Calf Serum). $5x10^4$ MKN-cells were cultured in 3.5cm culture dishes, at 37°C, in 5% CO_2. After 48 hours (day2), MKN-cells were washed with PBS and exposed to JCOE at various concentrations (100, 250, 500, 1000µg/ml), for 72 hours in medium. On day5 viable cells were counted by the Trypan blue dye exclusion test, and morphological changes were investigated with phase contrast microscopy.

Microscopic analysis for death cells   The effect of JCOE on apoptosis and necrosis was morphologically determined by fluorescent microscopic analysis. The cells were labeled with Hoechst 33342 (HO), propidium iodide (PI), and with two specific DNA-binding fluorescent dyes, as described by Nakajima et al. (6). In general, viable cells exclude PI, but are permeable to HO. Cells with a destroyed membrane are permitted to be stained by PI. Therefore, simultaneous staining with HO and PI makes it possible to discriminate whether the cells are alive or dead; necrotic, early- and late-apoptotic. $5x10^4$ MKN-cells were cultured in 3.5cm culture dishes, at 37°C in 5% CO_2. After 48 hours (day2) MKN cells were washed with PBS,and exposed to JCOE at various concentrations (500, 1000µg/ml) for 72 hours in medium. On day5, cells were incubated with 10 µg/ml of the HO dye for 15 min. at 37°C before the addition of 10 µg/ml of PI, for 10 min. Floating and attached cells were collected by trypsinization, washed in HBSS, and centrifuged (300 x g). The pellet was resuspended in 200µL of HBSS. 8µL of suspension was placed on a microscope slide, covered with a 22-mm^2 coverslip and examined under an x100 dry objective using epillumination. Apoptosis was morphologically characterized by the condensed chromatin of apoptotic cells.

FACS analysis 5×10^4 MKN-cells were suspended in medium, and cultured in 3.5cm culture dishes at 37°C, in 5% CO2. After 48 hours (day2) MKN cells were washed with PBS, and exposed to JCOE at various concentrations (500, 1000µg/ml) for 24 hours in medium. Cells were removed from the culture dishes by trypsinization, and were centrifuged. Following the fixation by 70% ethanol, cells were rehydrated in HBSS, centrifuged, and resuspended in 40 µl of phosphate-citric acid buffer containing 192 parts of 0.2 M Na2HPO4, and 8 parts of 0.1 M citric acid (pH 7.8). This was done at room temperature for 30 min., to facilitate removal of low molecular weight DNA from apoptotic cells, as described previously (7). The suspension was filtered through a 50 µm nylon mesh, and 50µg/ml propidium iodide (Calbiochem Corporation, La Jolla, CA) and 0.1mg/ml RNAse (Sigma Chemicals, St Louis, MO) were added. DNA content in stained nuclei was analyzed with FACScan (Becton Dickinson, Rutherford), and the number of stained nuclei in each phase was measured by the FACScan S-fit program (Becton Dickinson).

Results

JCOE Inhibited Cell Growth of MKN-45 Cells.

Figure 1 shows the growth curve of MKN cells after 72 hours incubation at various concentrations of JCOE. JCOE exerted cell growth inhibition of MKN-45 in a dose dependent manner within the concentration of 250-1000µg/ml, but did not affect MKN-28 cell growth. Although MKN-45 cells, after 72 hours incubation with JCOE (1000µg/ml) increased in number only about 20% as compared with control and the MKN-28 group. These cells that remained were alive, and their viability was not affected by JCOE after 72 hours exposure on these concentrations (data not shown). This means that JCOE inhibits only MKN-45 cell growth without affecting cytotoxicity.

Phase Contrast Macroscopic Findings of Both MKN Cells and Effect of JCOE on These Cells.

In the phase-contrast microscopic findings (x100), MKN-45 cells were growing in a colony-like formation, and the MKN-28 cells were forming a monolayer appearing like cobble stones. Although the MKN-28 cells increased in number without changing its cell-growth forms, MKN-45 cells decreased in number as compared with control with changing the number of colony formations after JCOE exposure (Figure 2). We could not find the difference between each MKN-45 cell form between the control and JCOE group after 72 hours incubation.

JCOE Induced Apoptosis in MKN-45 Cells.

In the fluorescent microscopic findings of the viable MKN-45 cells with HO and PI dye, cell nuclei was dyed blue with HO, but not dyed with PI. In the control group,

Table Ⅰ. The ingredients of *Crassosterera gigas* extract powder(JCOE)

Protein	23.50%
Amino acid	
Glutamic acid	2.78%
Proline	1.33%
Alanine	1.23%
Aspartic acid	1.20%
Glycine	1.10%
Lysine	0.54%
Arginine	0.53%
Threonine	0.51%
Leucine	0.45%
Serine	0.42%
Valine	0.37%
Phenylalanine	0.30%
Isoleucine	0.30%
Histidine	0.28%
Taurine	5.11%
Suggar	58.40%
Lipid	0.20%
Water	1.30%
Mineral,Vitamine	

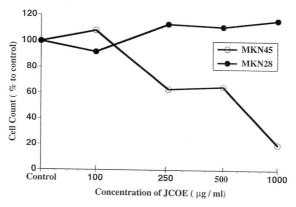

Figure 1
Growth curve of MKN cells after 72 hrs incubation at various concentrations of JCOE is shown. A number of 5×10^4 cells in a dish were cultured at 37Åé in 5% CO_2 for 48 hrs. Then MKN cells were exposed to various concentrations of JCOE (100, 250, 500,1000µg/ml) for 72 hrs. Cell number was counted by Trypan blue dye exclusion method. JCOE (250-1000 µg/ml) inhibited MKN-45 cell growth in a dose dependent manner, but did not affect the MKN-28 cell growth.

the cell nuclei were round and dyed blue. On the other hand, in JCOE group (500, 1000mg/ml), a small number of apoptotic cells, characterized by nuclear chromatin consendation and fragmentation, were seen. Apoptotic cells were often seen at a concentration of 1000mg/ml, but necrotic cells dyed red with PI were not seen (Figure 3). This data indicates that JCOE exposure to MKN-45 cells, induces its death in small numbers by apoptosis but not necrosis.

JCOE Increased S Peak on MKN-45 Cell-Cycle Progression.

To investigate the other mechanisms of the inhibition of MKN cell growth induced by JCOE, we tried to analyze its effect on the progression of the cell cycle. The phase distribution of the cell cycle of MKN cells was measured by flow-cytometric analysis with PI-stain. Exposure of JCOE (1000μg/ml) to MKN-45 cells resulted in a significant increase in the proportion of cells in the S peak of the DNA histogram; control: 42%, JCOE: 52% (Figure 4). Its proportion increased in a dose-dependent manner (data not shown). In the case of MKN-28, the S peak proportion was not changed by JCOE.

Discussion

Many natural products have been widely used in Japan as well as around the world, but their active ingredients and mechanism of action are minutely known. We have reported that some natural products, including Kampo medicines, scavenge free radicals, and that their actions may have important roles in the prevention of free radical injury (8). In our recent study, we have found the anti-oxidative properties in JCOE, and its cellular protection against reactive oxygen metabolites.

In the present study, we have investigated the effects of JCOE on cell growth in gastric carcinoma cell lines (MKN cells). JCOE inhibited MKN-45 cells growth in a dose-dependent manner , but not MKN-28 cells. The inhibition of MKN-45 cell growth was induced by cell cycle arrest, and/or apoptosis, but not by necrosis. The main difference between these MKN cell lines is thought to be based on a character of p53. It has been widely reported that p53 is a component of intracellular protein, composed of 393 amino-acids, and has some following functions; 1) transfer regulation, 2) intracellular signal transaction , 3) a component of protein-complex of DNA replication, 4) exonuclase activity. This results in the following events; 1) cell cycle arrest, 2) induction of appoptosis, 3) repair of DNA, 4) regulation of DNA replication, 5) induction of cell differentiation (9). It is also recognized that p53 is composed of two different types; a wild type and a mutant type. The latter is contained in about 50 % of all malignant tumor cells. MKN-45 has a wild p53, and MKN-28 , a mutant p53 (10).

Our data shows that JCOE did not affect the cell growth of MKN-28, which contained a mutant p53, but paused the cell growth of MKN-45, containing a wild p53. These results may implicate that JCOE activates a wild p53, but not a mutant p53 under certain conditions, resulting in cell cycle arrest and induction of appoptosis. We think that further study will be necessary to define the role of p53 in

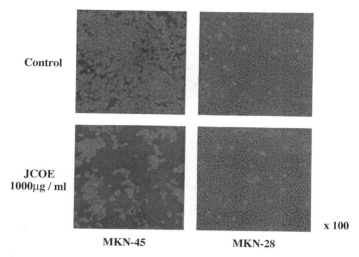

MKN-45   **MKN-28**

Figure 2
Effect of JCOE on phase contrast microscopic findings of MKN cells is shown.
Phase-contrast microscopic findings shows JCOE (1000μg/ml) inhibited MKN-45
cell growth but did not MKN-28 cell growth.

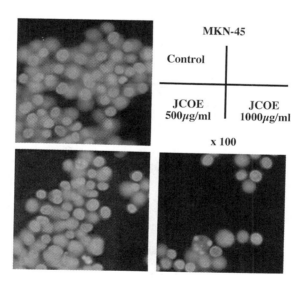

Figure 3
Effect of JCOE on induction of cell death which was apoptosis or necrosis was
morphologically determined by fluorescent microscopy after labeling with Hoechst
33342 (HO) and propidium iodide (PI). A little number of apoptotic cells
characterized by nuclear chromatin consendation and fragmentation were seen at
concentrations of JCOE (500, 1000mg/ml) on MKN-45 cells.

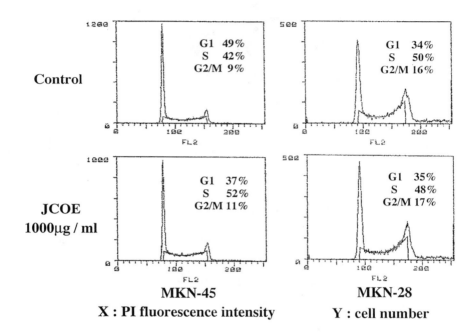

X : PI fluorescence intensity   **Y : cell number**

Figure 4

DNA content in stained nuclei with PI was analyzed with FACScan. Exposure of JCOE (100, 500, 1000μg/ml) on MKN-45 cells resulted in a significant increase in S peak of the DNA histogram in a dose-dependent manner but not on MKN-28 cells.

these MKN cells, and the mechanism of JCOEs actions on the interactions between the function of p53 and MKN cell growth inhibition.

In summary, treatment with JCOE inhibited growth in gastric carcinoma cell lines in a concentration dependent manner. These results indicate that JCOE is one of candidates for cancer-chemoprevention.

References

1. Ip, C.; Lisk, D. J.: Bioavailability of selenium from selenium-enriched garric. Nutr-Cancer. 1993; 20 (2): 129-137.
2. Zeng, G. Q.; Kenney, P. M.; Zhang, J.; Lam, L. K.: Chemoprevention of benzo[a]pyrene-induced forestmach cancer in mice by natural phthalides from celery seed oil. Nutr. Cancer. 1993; 19 (1): 77-86.
3. Azuine, M. A.; Bhide, S. V.: Aduvant chemoprevention of experimental cancer: catechin and dietaryturmeric in forestomach and oral cancer models. J Ethnopharmacol. 1994; 44 (3): 211-217.
4. Ip, C.; Lisk, D. J.: Bioactivity of selenium from Brazil nut for cancer prevention and selenoenzyme maintenance. Nutr. Cancer. 1994; 21 (3): 203-212.
5. Yoshikawa, T.; Naito, Y.; Masui, Y., et al.: Free radical-scavenging activity of Crassostera gigas extract (JCOE). Biomed &Pharmacother. 1997; 51.
6. Nakajima, H.; Park, H.L.; Henkart, P. A.: Synergistic roles of granzymes A and B in mediating target cell death by rat basophilic leukemia/mast cell tumors also expressing cytolysin/perforin. *J. Exp. Med.* 1995; 181: 1037-1046.
7. Gong, J.; Traganos, F.; Darzynkiewicz, Z.: A selective procedure for DNA extraction from apoptotic cells applicable for gel electrophoresis and flow cytometry. *Anal. Biochem.* 1994; 218: 314-319.
8. Yoshikawa, T.; Takahashi, S.; Ichikawa, H. *et al.*: Effects of TJ-35(Sugyaku-san) on gastric mucosal injury induced by -reperfusion and its oxygen-derived free radical-scavenging activities. J. Clin. Biochem. Nutr. 1991; 10: 189-196.
9. Saya, H.: Trouble shooter for physical crises. Cell Technology. 1997; 16 (4): 512-517.
10. Matozaki, T. et al.: Missense mutations and a deletion of the p53 gene in human gastric cancer. Biochem. Biophis. Res. Commun. 1992; 182: 215-223.

Chapter 7

Marine Plasmalogen and its Antioxidative Effect Intake of Dietary Fish Oil into Membrane Lipid and Their Stability Against Oxidative Stress

Kozo Takama[1], Kazuaki Kikuchi[2], and Tetsuya Suzuki[1]

Faculty of Fisheries, Department of Marine Bioresources Chemistry, Laboratory of Food Wholesomeness, [1]Hokkaido University, Hakodate 041, Japan
[2]Yaizu Suisan Kagaku Company, Ogawa-Shinmachi, Yaizu, Shizuoka, Japan

Feeding marine lipids to rats increased n-3 PUFA level in myocardial phospholipids (MPL). Comparing PE subclasses (diacyl- and alkyl-PE), the PUFA content in PE-plasmalogen (PE-plasmalogen) was particularly high, suggesting the dietary PUFA was preferentially incorporated into PE-plasmalogen molecules. Levels of rat myocardial lipofuscin-like substances, and TBARS, in the marine lipid-fed group were lower than that of the lard-fed one. Those findings suggest antioxidative activity of PE-plasmalogen. The antioxidative effect of PE-subclasses was examined with Fe^{2+}-ascorbic acid system, by using multilamellar liposomes. Inserting PE-plasmalogen into the PC-liposomes prolonged the induction period. However, once peroxidation was induced, it rather accelerated the reaction. PE-plasmalogen in the liposome exhibited a protective effect on n-3 PUFA, only in the early stage of peroxidation. These results suggest that antioxidative effects of dietary marine lipids may be due to PE-plasmalogen.

1-Alk-1-enyl-2-acylglycerophospholipid (general name; plasmalogen) is one of the lipid classes comprising membrane lipids. Generally, the C-2 position of the acyl moiety of glycerophospholipid is predominantly occupied by polyunsaturated fatty acids (PUFA) (*1*), which is assembled by an enzyme, acyltransferase (*2,3*). An interesting fact is that plasmalogen is rich in the tissues which demand high oxygen consumption; e.g., brain, heart, skeletal muscles, gills (*4~7*). In order to examine whether marine origin plasmalogen participates in the stabilization of membrane phospholipids or not, we made following studies; i.e., (a) comparison of intramolecular localization of PUFA in the rat myocardial membrane phospholipids before and after feeding diets of different marine lipid compositions., and (b) possible role of marine lipid origin plasmalogen as an antioxidant against oxidative damage of liposomes, as a biomembrane model.

Dietary Lipid and Its Reflection on The Fatty Acid Compositions of Lipid Subclasses in Rat Heart

Effect of Fish Oil Diet on Rat Heart Muscle Phospholipids. Two different types of diets were fed to male Wistar rats (5 weeks after weaning), for 4 weeks. The diets used in this study were: 1.) a soybean diet, consisting AIN-76 basal diet, and 5% of soybean oil (Asahi-Yushi KK. Asahikawa, Hokkaido), and 2.) a fish oil diet which was composed of AIN-76 diet and 4% of soybean oil and 1% of sardine oil (Nihon Kagaku Shiryo, Hakodate, Hokkaido). After 4 weeks of feeding, the rats were fasted overnight, then euthanized under diethyl ether anesthesia by depleting whole blood. Next, the heart was taken out and washed thoroughly with PBS-saline to remove the blood. Hearts thus prepared were provided for lipid analysis. Heart lipid extracts were fractionated into phosphatidylcholine (PC), and phosphatidylethanolamine (PE) on a preparative TLC according to Horrocks et al. (8). These phospholipid fractions were further subfractionated into diacyl-type, alkenyl-type and alkyl-type according to Waku et al. (9). They were provided for the analysis of fatty acid composition.

As shown in Table I the proportion of n-3 series PUFA, represented by 20:5 and 22:6 in PE of any lipid subclasses, was higher than those of PC. The n-3 PUFA was remarkably high in alkenyl type PE (PE-plasmalogen) (Figure 1). It has been well established that the fatty acid composition of animal cell membranes is strongly influenced by dietary fat composition (10). Among them, the incorporation efficiency of PUFA into plasmalogen, especially into PE-plasmalogen was found extraordinary high.

Comparison of lard diet, fish oil diet, and marine phospholipid diet on the fatty acid composition of heart muscle membrane lipid and TBARS levels. In another experiment, we fed rats three different types of diets; a) First, with a 1% of lard (Tsukishima Food Ind. Corp., Tokyo, Japan), b) secondly, fish oil (Nihon Kagakushiryo, Hakodate, Hokkaido, Japan), c) and thirdly, marine phospholipids were separately added to AIN-76 diet. The marine phospholipid fraction was prepared in our laboratory by extracting the crude lipid fraction with cold acetone from sardine gonads. Feeding was continued for 4 weeks, to male Wistar rats (6 weeks after weaning). Results showed that the fatty acid composition of myocardial lipids was closely dependent upon the fatty acid composition of dietary lipids (Table II). Lard diet fed group showed the highest TBARS level (11), whereas that of marine phospholipid diet fed group was the lowest (Table III).

Comparison of Lipofuscin Levels of Rat Liver Muscle

Lipofuscin is an amyloid complex formed by the reaction of lipid peroxide, amino acids, protein, and nucleic acids (12, 13). Lipofuscin has been regarded as an index of senescence ongoing because it accumulates in animal tissues accompanied with aging. A lipofuscin like substance is known to give bright yellow color under the fluorescence microscope. Heart muscle sections prepared by formalin fixation, and subsequent paraffin embedding

Table I. Fatty Acid Composition (%) of PC- and PE- Subclasses in the Rat Heart.

Fatty Acid	Diacyl PC SO	FO	PE SO	FO	Alkenyl PC SO	FO	PE SO	FO	Alkyl PC SO	FO	PE SO	FO
16 : 0	19.6	21.1	10.1	11.5	9.9	10.1	1.9	1.1	25.4	28.9	17.5	18.3
16 : 1	0.3	0.4	0.1	0.1	1.3	1.9	0.5	0.6	2.2	4.8	2.3	3.0
18 : 0	25.1	24.5	30.7	31.2	4.7	4.5	2.8	2.0	6.3	7.0	7.7	6.5
18 : 2	10.9	4.6	6.4	2.2	6.8	4.9	2.4	1.0	10.2	8.8	10.1	6.3
20 : 4	27.5	21.5	17.4	8.4	49.1	35.1	43.9	20.3	25.2	19.3	13.5	10.9
20 : 5	0.2	3.1	0.1	1.8	1.2	3.5	0.5	5.5	1.1	3.6	0.7	6.1
22 : 4	0.8	0.1	1.1	0.1	1.5	tr	4.2	0.4	1.1	tr	4.1	0.9
22 : 5	1.7	3.5	2.5	2.1	3.3	4.7	6.6	9.4	1.1	3.1	12.4	11.7
22 : 6	4.7	13.2	23.8	37.8	12.9	27.5	29.5	53.4	7.3	10.8	17.8	24.0
Σ Sat	44.8	45.6	40.8	42.7	17.9	17.0	7.4	6.0	43.2	42.5	29.0	30.5
Σ PUFA	46.2	46.2	53.3	53.2	76.1	76.6	89.2	91.4	48.3	45.7	60.9	61.0
Σ n-6	39.6	26.4	26.9	11.5	58.6	40.9	52.5	23.0	38.8	28.1	29.9	19.3
Σ n-3	6.6	19.8	26.4	41.7	17.4	35.7	36.6	68.3	9.5	17.6	31.0	41.7

PC: Phosphatidylcholine, PE: Phosphatidylethanolamine, SO: Soybean Oil, FO: Fish Oil . PC contents (μ mol/ g heart) : SO fed group 12.5±0.7 [Diacyl 10.3±1.5, Alkenyl 0.5±0.3, Alkyl 0.2±0.2] ; FO fed group 13.0±0.9 [Diacyl 12.3±2.0, Alkenyl 0.8±0.1, Alkyl 0.3±0.3], PE contents (μ mol/g heart) : SO fed group 11.2±0.9 [Diacyl 7.6±0.7, Alkenyl 3.1±0.5, Alkyl 0.2±0.1] ; FO fed group 11.7±1.2 [Diacyl 7.9±0.4, Alkenyl 3.1±0.1, Alkyl 0.2±0.1].

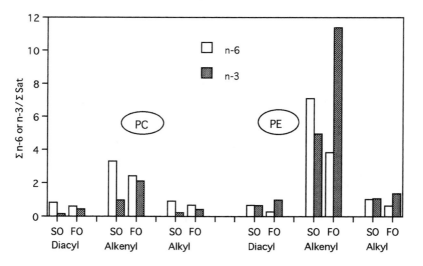

Figure 1. Relative distributions of PUFA(n-6 or n-3) against total saturated fatty acid.

□: n-6, ■: n-3 ; SO: Soybean oil, FO: Fish oil

Table II. Fatty Acid Composition (%) of Lipids in the Diet and Heart of the Rats after Feeding for Four Weeks.

Fatty Acid	Diet	NPL	PL	Diet	NPL	PL	Diet	NPL	PL
14 : 0	0.4	1.9	0.1	1.6	0.8	0.2	1.1	0.7	1.2
16 : 0	13.1	25.5	17.6	15.3	12.8	15.9	22.9	16.3	14.6
16 : 1	0.4	8.0	0.2	1.5	2.1	0.1	0.4	2.3	0.2
18 : 0	4.9	3.1	32.6	3.5	2.9	31.4	3.0	5.4	30.4
18 : 1	18.4	29.5	3.0	22.1	18.5	2.3	24.8	22.2	2.6
18 : 2	45.5	29.8	12.2	45.3	55.4	16.8	38.6	42.7	14.6
18 : 3	6.6	0.8	-	5.6	0.7	0.1	4.8	0.8	0.1
20 : 4	1.2	1.1	22.2	0.8	2.4	18.2	0.3	0.3	16.1
20 : 5	-	-	-	2.4	-	0.4	0.4	0.4	0.1
22 : 6	-	0.7	11.3	1.9	2.4	13.5	3.7	3.7	18.5

NPL: non-polar lipids, PL: polar lipids.

Table III. Induction Period and Peroxidation Rate of PC-Liposome in Fe^{2+}-Ascorbate System.

Tested lipid inserted into PC-liposome	PC	PE	PE-plasmalogen
Induction Period	0.098	0.166	0.226
Oxidation Rate (nmol MDA/hr)	26.3	33.7	40.9
Oxidation level (nmol MDA/mg lipid after 10hr)	51.2	60.0	71.2

PC: Phosphatidycholine, PE: Phosphatidylethanolamine

were examined under the fluorescence microscope equipped with Video Enhanced Image Analyzer (ARGUS-100, Hamamatsu Photonics, KK., Hamamatsu, Japan). Integrated fluorescence intensities of heart muscle sections are presented in Table III. Undoubtedly, the lard diet fed section gave the highest fluorescence level, and the lowest level was observed for the marine phospholipid diet fed section. The results presented above are definite evidence to prove acyl moieties of dietary lipids. The n-3 series PUFA are especially incorporated into the acyl moieties of PE -plasmalogen. Furthermore, a very interesting result, was that lipid hydroperoxide level was kept low in the samples rich in n-3 series PUFA contents, because n-3 series PUFA has been regarded highly susceptible to reactive oxygen attack.

Susceptibility of Lipid Peroxidation and Participation of PE-plasmalogen in Liposomal System

The results mentioned above suggest a possible relationship between PE-plasmalogen and suppression of membrane lipid peroxidation. Therefore, we carried out model experiments with an iron-ascorbic acid system, in diacyl-PE and alkenyl-PE embedded multilamellar liposomes, to examine the possibility whether our postulation would be worthy or not. PE-plasmalogen embedded multilamellar liposomes were prepared from egg yolk phosphatidylcholine (Sigma Chemical Co., St. Louis, MO) and 10% (mol/mol) PE-plasmalogen, by the method of Nagao and Mihara (14), with slight modification. In order to induce lipid peroxidation in the liposomes, egg yolk PC-OOH (15,16) was added at the concentration of 0.005% (mol/mol). Iron-ascorbic acid-induced lipid peroxidation was carried out in HEPES buffer (pH 7.4), at 37 °C At regular time intervals, certain volumes of reactant were collected for the measurement of TBARS, by the method of Uchiyama and Mihara (17).

A relationship between the malondialdehyde (MDA) concentration, and incubation time, was plotted on a semi-logarithmic scale as shown in Figure 2. Since a linear relation was obtained in every experimental condition, the distance to the right of 0 along the x-axis, where lines intersects in the x-distance, should be regarded as the induction period. On the other hand, the gradient of each line is regarded to mean the rate of oxidation velocity. Obviously, the induction period was more effectively elongated in alkenyl-PE embedded liposomes than PE embedded system. However, once the lipid peroxidation was started, it was rather accelerated in the presence of plasmalogen. Major PUFA, consisting of PE (prepared from porcine liver, Doosan Serdary Res. Lab. Inc., Englewood Clifs, NJ) used in the present experiment, were linoleic acid (18:2) and arachidonic acid (20:4), occupying 9% and 16%, respectively. The ratio of PUFA to the total sum of saturated fatty acid was 0.9. While PE-plasmalogen (prepared from bovine brain, Doosan Serdary Res. Lab. Inc., Englewood Clifs, NJ) had linoleic acid (18:2) and cis-7,10,13,16-docosatetraenoic acid (22:4) as the major PUFA occupying 29% and 9%, respectively, and the ratio of PUFA to total saturated fatty acids was 1.9. Thus the composition of fatty acids varied by the materials.

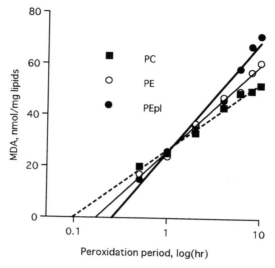

Figure 2. Peroxidation of PC-liposomes under Fe^{2+}-ascorbate system at 37℃.

■: Phosphatidylcholine, ○: Phosphatidylethanolamine,
●: PE-plasmalogen

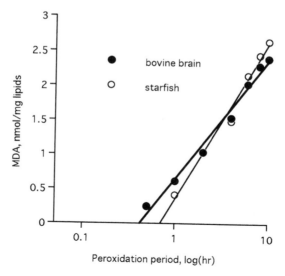

Figure 3. PC-liposome peroxidation induced by AAPH at 37℃.

●: bovine brain PE-plasmalogen, ○: starfish PE-plasmalogen

64

Extension of initiation stage by PE-plasmalogen

Although it is still unclear whether the compositional differences of fatty acids was the key to the antioxidative effect of PE-plasmalogen or not, an experiment using liposomes embedding PE-plasmalogen, which was prepared from starfish (Astropecten scoparius), showed us quite interesting results. As much as 74.7% of the PE fraction of the starfish was composed of alkenyl-type, or PE-plasmalogen. In this model experiment, peroxidation was started by the addition of an water soluble radical initiator, AAPH [2,2'-azobis(2-amidinopropane)dihydrochloride]. In comparison with the previous experiment using bovine brain plasmalogen, the starfish-origin plasmalogen clearly showed a better effect on the extension of the induction period. At the same time, the rate and the extent of oxidation was also enhanced by the starfish-origin PE-plasmalogen than the bovine brain one. The starfish PE-plasmalogen contained arachidonic acid (34%) and eicosapentaenoic acid (20%), and the ratio of sum of PUFA to the total saturated fatty acid was 4.4. It is likely that higher content of PUFA in plasmalogen suppresses initiation of lipid peroxidation, or extends induction period. But once the peroxidation reaction moves into the propagation stage, PE-plasmalogen with higher PUFA content accelerates oxidation velocity.

Antioxidative effects at the induction stage and a stimulation effect at the propagation stage of the PE-plasmalogen should be derived from the vinyl-ether bond at C-1 position of glycerol moiety. Kobayashi (18) commented in his review article the participation of vinyl-ether linkage during u.v.-induced oxidation reaction; i.e., continual U.V. irradiation leads to decomposition of plasmalogen in the liposomal system with time, and its decomposition is characteristic to the compound with a vinyl-ether linkage. That is, during peroxidation reaction in the liposomes plasmalogen molecules, or vinyl-ether bond, was served preferentially to be oxidized and decomposed by the singlet oxygen. In other reactions, the reaction mechanism with other reactive oxygen and free radicals should be the same. This is because induction periods were extended both in the AAPH-induced, and hydroperoxide-initiated systems.

Furthermore, plasmalogen has been reported to participate in the protection of cell function of the Chinese hamster ovary (CHO) cells (19). Its protective function is explained by the high susceptibility of the molecule to reactive oxygen attack (20). In our experiment, a decrease of PUFA was recognized along with promoting propagation reaction in the liposome system (data not shown). In brief, plasmalogen serves itself as a sacrifice to protect other membrane lipid components.

Literature Cited.

(1) Horrocks, L.A. In *Ether Lipids, Chemistry and Biology*; Snyder, F. Ed.: Content, composition, and metabolism of mammalian and avian lipid that contain ether groups; Academic Press: New York, NY, **1972**, pp 233-237.
(2) Svennerholm, L.; Stallberg-Stenhagen, S. *J. Lipid Res.* **1968**, *9*, 215-225.
(3) Segler-Stahl, K.; Webster, J.C.; Brungraber, E.C. *Gerontology.* **1983**, *29*, 161-168.

(4) Horrocks, L.A. In *Ether Lipids, Chemistry and Biology*; Snyder, F. Ed.: Content, composition, and metabolism of mammalian and avian lipid that contain ether groups; Academic Press: New York, NY, 1972, pp 179-224.

(5) Davenport, J.B. *Biochem. J.* **1964**, *90*, 116-122.

(6) Spanner, S. *Nature*, **1966**, *210*, 637.

(7) Chapelle, S. *Comp. Biochem. Physiol.* **1986**, *85B*, 507-510.

(8) Horrocks, L.A.; Sun, G.Y. In *Research Methods in Neurochemistry*, Rodnight, R.; Mariks, N. Ed.; Ehanolamine plasmalogens; Plenum Press: New York, NY, 1972; pp 223-231.

(9) Waku, K.; Ito, H.; Bito,T.; Nakazawa, Y. *J. Biochem.* **1974**, *75*, 1307-1312.

(10) Scott, T.W.; Ashes, J.R.; Fleck, E.; Gulati, S.K. *J. Lipid Res.* **1993**, *34*, 827-835.

(11) Fukunaga, K.; Suzuki, T.; Takama, K. *J. Chromatogr.* **1993**, *621*, 71-81.

(12) Chio, K.S.; Reiss, U.; Fletcher, B.; Tappel, A.L. *Science.* **1969,** *166*, 1535-1536.

(13) Kikugawa, K.; Ido, Y. *Lipids.* **1984**, *19*, 600-608.

(14) Nagao, A.; Mihara, M. *Biochem. Biophys. Res. Commun.* **1978**, *172*, 385-389.

(15) Terao, J.; Matsushita, S. *Agric. Biol. Chem.* **1981**, *45*, 587-593.

(16) Terao, J.; Hirota, Y.; Kawakatsu, M.; Matsushita, S. *Lipids.* **1981**, *16*, 427-432.

(17) Uchiyama, M.; Mihara, M. *Anal. Biochem.* **1978**, *86*, 271-278.

(18) Kobayashi, T. In *Function and Biosynthesis of Plasmalogens*; Inoue, K., Nozawa, Y., Ed.; Phospholipid metabolism and disease in animal cells published as an extra number of Protein, Nucleic Acid and Enzyme; Kyoritsu Shuppan Kabushiki Kaisha: Tokyo, 1991, Vol. 2; pp 424-432. (in Japanese)

(19) Zoeller, R.A.; Morand, O.H.; Raetz, C.R.H. *J. Biol. Chem.* **1988**, *263*, 11590-11596.

(20) Morand, O.H.; Zoeller, R.A.; Raetz, C.R.H. *J. Biol. Chem.* **1988**, *263*, 11597-11606.

Processed Foods

Chapter 8

Antimutagenic Mechanism of Flavonoids Against a Food-Derived Carcinogen, Trp-P-2, Elucidated with the Structure–Activity Relationships

K. Kanazawa, H. Ashida, and G. Danno

Department of Biofunctional Chemistry, Faculty of Agriculture, Kobe University, Rokkodai, Nada-ku, Kobe 657, Japan

This study found a mechanism of the strong antimutagenicity of flavonoids against a food-derived carcinogen, 3-amino-1-methyl-5H-pyrido[4,3-b]indole (Trp-P-2). Trp-P-2, after being incorporated into the cells, is metabolically activated by cytochrome P450 (P450) 1A family to 3-hydroxyamino-Trp-P-2, as the ultimate carcinogen. Flavonoids inhibited the N-hydroxylation process, showing the structure-activity relationships. Their inhibitory IC_{50}s coincided with the antimutagenic IC_{50}s, which were determined by the *Salmonella* test. Particularly, flavones and flavonols were the strong inhibitors, as their Ki values on P450 1A1 were less than 1 μM, and very low comparing to the Km value of Trp-P-2 (25 μM). Thus, the antimutagenicity of flavonoids against Trp-P-2 was due to the high affinity to P450 enzyme, which was a preventing effect from the initiation step in dietary carcinogenesis.

We have found the strong, and specific antimutagens against the food-derived carcinogen, 3-amino-1-methyl-5H-pyrido[4,3-b]indole (Trp-P-2) (1-3). The active compounds are flavonoids such as apigenin, galangin and luteolin that widely spread in our daily vegetables (4). Trp-P-2 is one of carcinogenic heterocyclic amines that are formed during daily cooking processes (5). For example, a dried sardine, when grilled, mildly produces 13.1 ng/g of Trp-P-2, 13.3 ng/g of 3-amino-1,4-dimethyl-5H-pyrido[4,3-b]indole (Trp-P-1), and small amounts of others (6). These heterocyclic amines are considered to be one of the major causes in human cancer (7, 8), because humans have been estimated to ingest them with 0.4-16 μg per day, per capita (9). Among the heterocyclic amines, Trp-P-2 is considered to be the strongest mutagen and carcinogen (10). Therefore, understanding the antimutagenic mechanism of flavonoids is an important issue for our health care.

Flavonoids inhibited the mutagenicity of Trp-P-2 with a dose-dependent manner, and the activity was great beyond comparison with other phytochemicals; chlorophylls, vitamins, carotenoids, xanthophylls, sterols, and saponins (2). On the mechanism of the antimutagenicity, there are three assumptions; one is a contribution by the antioxidative potency of flavonoids (11-14). Second is an inducing effect of G2/M arrest on the mutated cells as flavones and isoflavones have been found to be the reversible inducer (15, 16). Another is an inhibition of cytochrome P450 (P450) enzymes that metabolically activate Trp-P-2 to its ultimate carcinogenic form (17-20). This study finished the discussion providing three evidences; the antimutagenicity of flavonoids was apparent, regardless of the antioxidative potency, flavonoids were a desmutagen to neutralize Trp-P-2 before attacked DNA, and flavonoids, particularly flavones and flavonols, acted as a specific inhibitors on P450 1A1.

Materials and Methods

Chemicals. Trp-P-2 was purchased from Wako Pure Chem. Ind., Ltd. Flavonoids were used in high grade analyzing with nuclear magnetic resonance (NMR) spectra (Bruker AC-250) after obtaining from Extrasynthèse S.A.Co., Wako Pure Chem. Ind. Ltd, Nacalai Tesque, Inc, Aldrich Chem. Co. Inc., Tokyo Kasei Kogyo Co. Ltd, Funakoshi Co. Ltd., and Kurita Kogyo Co. Ltd. Organic solvent and water were distilled twice, and all other chemicals were of the highest grade available from commercial sources.

The hydroxyl-masked chemicals with methyl or acetyl were prepared from luteolin and quercetin. They were methylated under diazomethane gas 3 times, or acetylated in a mixed solvent of 1.5 ml anhydrous pyridine and 0.75 ml acetic anhydride at 25 °C for 11 h. These products were separated on a silica gel column, and then purified by a recrystallization. The NMR spectra showed that all of hydroxyl groups on luteolin and quercetin were blocked with methyl or acetyl.

Quantitative Comparison in the Antimutagenicity of flavonoids. The antimutagenicity of flavonoids against 0.1 nmol Trp-P-2 was evaluated with *Salmonella typhimurium* TA98 strain in the presence of S9 mix (containing 0.11 nmol P450 enzymes) as mentioned previously (21). Six-different concentrations of flavonoids in dimethylsulfoxide (0.1 ml) were mixed with Trp-P-2 in water (0.1 ml), and incubated with 0.5 ml S9 mix and bacterial suspension (0.1 ml) at 37°C for 20 min. The mixture was added to 2 ml molten top agar, and poured onto an agar medium of minimal glucose. Culturing for 2 days, the numbers of His[+] revertant colonies were counted.

The desmutagenicity data was calculated as follows: $\{(A-B)-(C-D)\}/(A-B) \times 100$; A, revertant number given by 0.1 nmol Trp-P-2; B, spontaneous revertants; C, revertants by both flavonoids and 0.1 nmol Trp-P-2; and D, revertants by each flavonoid. When the six-different concentrations of flavonoids constructed a dose-response curve, the IC_{50} values, the amount to require for 50% inhibition of the mutagenicity of 0.1 nmol Trp-P-2, were determined as shown in Figure 1.

Determination of Inhibitory Effects of Flavonoids on P450 1A1. We received from Prof. Ohkawa (Kobe University, Japan) a kind gift of *Saccharomyces cerevisiae* AH22/pAMR2 cells, expressing both rat P450 1A1 and yeast reductase simultaneously (*22*). The strain was cultured monitoring an increase of the P450 content in the cell, in a medium composed of 8% glucose, 5.4% yeast nitrogen base w/o amino acids (Difco Laboratory), and 0.016% histidine for 3 days. They were then submitted to the following microsomal preparation (*23*). The cell suspension in the Tris-HCl buffer (200 ml) was mixed with 60 mg of zymolyase, incubating at 30°C for 1 h. After the centrifugation, the cell precipitate was resuspended in Tris-HCl buffer containing 0.65 M sorbitol, 0.1 mM dithiothreitol, 0.1 mM ethylenediaminetetraacetate, and 1 mM phenylmethylsulfonylfluoride, and then sonicated for 1 min twice. The microsomal fraction was obtained by a centrifugation at 105,000 x g for 70 min of the supernatant from 10,000 x g for 10 min, and resuspended in 10 mM of potassium phosphate buffer (pH 7.4) containing 20% glycerol, and 1 mM phenylmethylsulfonylfluoride. The P450 content was determined by the reduced CO-difference spectrum when the microsomal fraction was treated with CO-gas bubbling, and sodium hydrosulfite. The protein content in the microsomal fraction was measured by the method of Bradford (*24*).

The microsomes (0.23 mg as protein and 0.052 nmol as P450) were mixed with 20 nmol Trp-P-2 (excess amount) in 0.5 ml of 0.1 mM sodium phosphate buffer (pH 7.0) containing 0.3 mM NADPH. After 3-min incubation at 37°C, the N-hydroxylation activity of microsomes was determined with the produced 3-hydroxyamino-Trp-P-2 (N-hydroxy-Trp-P-2) as mentioned previously (*25*). Immediately adding 0.5 ml acetonitrile to the incubation mixture and centrifuging at 3500 rpm for 3 min, the supernatant (10 μl) was analyzed by a high-performance liquid chromatograph (Shimadzu LC-6AD), equipped with an electrochemical detector (IRICA Σ875) at +300 mV. A column, Inertsil ODS (ø 4.6 x 150 mm), was maintaining at 40°C, eluting with 20 mM potassium phosphate monobasic (pH to 4.6)/ acetonitrile (80/20, v/v) containing 0.1 mM ethylenediaminetetraacetate disodium salt at the flow rate of 1.0 ml/min. This method can determine N-hydroxy-Trp-P-2 to the levels up to 1 pmol without interference with any impurities.

To determine the inhibitory effect of flavonoids on the N-hydroxylation, six-different concentrations of flavonoids were added to the incubation mixture. After the four independent determinations with three tests per each concentration of flavonoids, the IC_{50} values of flavonoids for the inhibition were calculated by

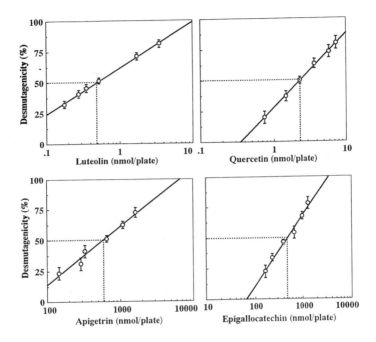

Figure 1. Determination of IC$_{50}$s for the desmutagenicity of flavonoids against Trp-P-2. Six-different concentrations of flavonoids were added to 0.1 nmol Trp-P-2 and incubated with S9 mix (containing 0.11 nmol P450 enzymes) and bacterial suspension at 37°C for 20 min. The mixture was cultivated and then the revertant colony number was counted as mentioned in the Text. After the four independent tests with three plates per each concentration of flavonoids, the desmutagenicity was plotted versus log of the dosed amount. The IC$_{50}$ values were determined from the amount to require for 50% inhibition of the mutagenicity of Trp-P-2.

plotting the produced N-hydroxy-Trp-P-2 versus log of the flavonoid concentrations.

Results

Antimutagenic Mechanism of Flavonoids. Trp-P-2, after being absorbed into the body (26), is metabolically activated by hepatic P450 monooxygenases to the ultimately carcinogenic form, N-hydroxy-Trp-P-2 (27, 28). The N-hydroxy-Trp-P-2 easily generates its radical, and attacks DNA to induce a frame-shift mutation (29, 30), which is the initiation step in the carcinogenesis. Therefore, flavonoids are assumed to participate in any of the following five ways; (a) insolubilize Trp-P-2 before being incorporated into the cells, (b) inhibit the metabolic activation to N-hydroxy-Trp-P-2 by P450 enzymes, (c) neutralize N-hydroxy-Trp-P-2, (d) scavenge the radicals from N-hydroxy-Trp-P-2 before attack DNA, and/or (e) suppress the mutation fixation process after DNA has been damaged by N-hydroxy-Trp-P-2.

To examine the function, (a) flavonoids were mixed with Trp-P-2 in water and immediately centrifuged. Then, the remaining Trp-P-2 in the supernatant was determined via HPLC (data not shown). Luteolin and quercetin made the precipitate with Trp-P-2 in a dose-dependent manner, but chrysin did not show the dose-dependent manner, and galangin was non-active. To insolubilize 50% of Trp-P-2 in the solution, luteolin required the 140-fold amount toward Trp-P-2 and quercetin required 190-fold amount. The large amounts of flavonoids were greatly different from the required amounts for their antimutagenicity, as shown in our previous reports (2, 3). Thus, the antimutagenicity of flavonoids was not considered to be attributed to the function (a).

The questions (b) and (d) will be explained minutely in the following session. Table I examines the function (c) with galangin and quercetin. Trp-P-2 was previously activated to N-hydroxy-Trp-P-2 with S9 mix before the incubation with galangin or quercetin. Assaying the mutagenicity of the mixture, both flavonoids were almost non-effective on the revertant number given by N-hydroxy-Trp-P-2, even with two-order larger concentrations than the required amounts for the antimutagenicity against Trp-P-2 (2). Flavonoids also did not have the function (c).

Table II shows the effect of flavonoids on the previously mutated *Salmonella typhimurium* TA98 cells. The cells were mutated by Trp-P-2 in the presence of S9 mix (30), and then mixed with flavonoids. Both flavonoids decreased little the revertant number even with two-order larger amounts than the required amounts for

Table I.   The Effect of Galangin and Quercetin on the Mutagenicity of N-Hydroxy-Trp-P-2

Flavonoid	nmol/plate	revertants/plate[a]
Galangin	0	1147±79
	1.8	981±17
	7.4	1123±59
	37	1143±61
Quercetin	5.9	911±50
	30	1053±39
	148	927±46

[a]Trp-P-2 was metabolically activated to N-hydroxy-Trp-P-2 by the 20-min incubation at 37°C with the S9 mix.   After boiling for 1 min to inactivate the S9 enzymes, the mixture was added to the flavonoid solution and incubated again for 15 min at 37°C, and then submitted to the test with *Salmonella typhimurium* TA98 cells as mentioned in the Text.   The revertant number of the cells is represented as Mean±SD minus the spontaneous revertants (n=6).   SOURCE: Adapted from ref. 2.

Table II.   Bio-antimutagenicity Test of Galangin and Quercetin

Flavonoid	nmol/plate	revertants/plate[a]
Galangin	0	3156±156
	3.7	2874±116
	18	2954±129
Quercetin	14.8	2413±94
	59	2722±96

[a]*Salmonella typhimurium* TA98 cells were previously mutated by the 20-min incubation at 37°C with Trp-P-2 and S9 mix in the absence of flavonoids.   The cells were incubated again with the flavonoids for 20 min, and then assayed the revertant number, which is represented as Mean±SD minus the spontaneous revertants (n=6).   SOURCE: Adapted from ref. 2.

the antimutagenicity (2).   The results mean that flavonoids were non-effective on the mutation fixation process of the cells and that they did not have the function (e).

The antimutagenicity is classified into desmutagenicity and bio-antimutagenicity, according to modes of action (31).   The desmutagenicity implies the various activities to neutralize a mutagen before attacks DNA.   The bio-antimutagenicity means suppressing activity at the process of mutation fixation, after DNA has been damaged by a mutagen.   The above results indicate that flavonoids were a desmutagen to neutralize Trp-P-2 before or during the metabolic activation, but not a bio-antimutagen (*31*).

Structure-Activity Relationships of Flavonoids. When the desmutagenicity of flavonoids is due to (b), the inhibitory effects of flavonoids on P450 enzymes should be close to their desmutagenic potency against Trp-P-2. In the case of (d), the desmutagenicity should correlate with the hydroxyl number and position on flavonoid skeleton because have been well recognized to closely associate with the antioxidative potency (*14, 32-34*). Then, we determined the desmutagenic IC$_{50}$s of 32 flavonoids, and the derivatives to compare quantitatively with their inhibitory effect on the N-hydroxylation of Trp-P-2 by the P450 enzyme. This was also done to examine the function (d) based on their structure-activity relationships (Table III).

N-Hydroxylation of Trp-P-2 had been considered to be catalyzed by P450 1A1 and P450 1A2 monooxygenases (*28*). Then, we measured the activity of microsomes from the recombinant *Saccharomyces cerevisiae* AH22/pAMC1 cell expressing a rat P450 1A1 (*35*) and AH22/pACDD2 cell expressing a modified rat P450 1A2 (*36*) (data not shown). Determining the activity at the pre-steady state up to 3 min of the incubation, the former microsomes had a slightly higher activity than the latter. Since the P450 enzymes require a NADPH-dependent P450 reductase (*22*), we also measured the activity of AH22/pAMR2 cell, expressing both rat P450 1A1 and yeast reductase simultaneously (*22*). The microsomes from AH22/pAMR2 cell gave one-order higher activity than the microsomes from AH22/pAMC1 cell. Then, to examine the function (b), the microsomes from the AH22/pAMR2 cell were used in a system containing the excess amount of Trp-P-2 and with the 3-min incubation.

In Table III, the smaller IC$_{50}$ value means the greater activity. A typical desmutagenic phytochemical has been reported to be chlorophyll (*37-39*). The IC$_{50}$ of chlorophyll has been determined to be 260 nmol (*3*), and flavonoids tested here manifested the greater activity except for catechins and flavonoid glycosides. Flavonoids are classified into flavones, flavonols, flavanones, and isoflavones based on a feature of the chemical structure. Flavones and flavonols are different in the steric structure from flavanones and isoflavones. Flavones and flavonols are flat in the steric structure because have the unsaturated 2, 3-bond in the pyranone ring (C-ring). Contrary, flavanones have the saturated 2, 3-bond, and isoflavones bind the phenyl group (B-ring) on 3 position of C-ring. Flavones and flavonols gave the lower IC$_{50}$ values than eriodictyol of flavanones and isoflavones. Regarding hydroxyl number, flavone having no hydroxyl group gave the lowest IC$_{50}$ among the flavonoids tested here. Also, flavanones having no hydroxyl group showed stronger activity than other flavonoids having four or more hydroxyl groups. Thus, the steric structure seemed to be more important for the desmutagenicity of flavonoids than the hydroxyl number, associating with the antioxidative potency. Then, being based on the structure-activity relationships, the function (d) was first examined.

Table III. IC$_{50}$ Values of Flavonoids for the Desmutagenicity and for the Inhibition of N-Hydroxylation by P450 1A1

Flavonoids (hydroxy positions)	IC$_{50}$ (nmol)[a] for:	
	desmutagenicity[b]	P450 inhibition[c]
vone (2-phenyl-4H-1-benzopyran-4-one)		-Fla
Flavone (-)	0.23	0.11
Chrysin (5, 7)	2.0	0.20
Baicalein (5, 6, 7)	2.5	-
Apigenin (5, 7, 4')	0.88	0.075
Luteolin (5, 7, 3', 4')	0.49	0.11
Flavonol (3-hydroxyl flavone)		
Flavonol (3)	1.0	0.080
Galangin (3, 5, 7)	0.44	0.19
Kaempferol (3, 5, 7, 4')	0.88	0.11
Morin (3, 5, 7, 2', 4')	1.2	-
Quercetin (3, 5, 7, 3',4')	2.4	0.18
Myricetin (3, 5, 7, 3',4', 5')	2.5	0.20
Fisetin (3, 7, 3', 4')	3.8	-
Flavanone (2-3 is saturated bond)		
Flavanone (-)	1.6	2.5
Naringenin (5, 7, 4')	2.1	-
Eriodictyol (5, 7, 3', 4')	42	-
Isoflavone (3-phenyl)		
Genistein (5, 7, 4')	13	-
Daidzein (7, 4')	79	95
Hydroxy-masked Flavonoid		
Isorhamnetin (Quercetin-3'-methoxy)	0.32	-
Hesperetin (Eriodictyol-4'-methoxy)	14	-
All-Methoxyl luteolin	2.5	2.0
All-Methoxyl quercetin	3.2	2.0
All-Acetyl luteolin	0.96	-
All-Acetyl quercetin	7.1	-
Catechins (saturated benzopyran)		
Catechin (3, 5, 7. 3', 4')	non-effective at less than 552 nmol	
Epicatechin (3, 5, 7, 3', 4')	non-effective at less than 2800nmol	
Epigallocatechin (3, 5, 7, 3', 4', 5')	460	-
Epicatechin gallate (Epicatechin-3-O-galloylester)760		-
Flavonoid Glycosides		
Rutin (Quercetin-3-O-rutinoside)	30	20
Quercitrin (Quercetin-3-O-L-rhamnoside)	4.1	-
Apigetrin (Apigenin-7-O-glucoside)	590	non-effective
Daidzin (Daidzein-7-O-glucoside)	non-effective between 0.024 and 960 nmol	
Puerarin (Daidzein-8-rhamnoside)	non-effective between 0.024 and 960 nmol	

[a]The IC$_{50}$s. which are the amounts requiring 50% inhibition of the mutagenicity of 0.1 nmol Trp-P-2 or of the enzymatic N-hydroxylation. were calculated from four independent experiments with three tests per each concentration of flavonoids.

[b]The desmutagenic IC$_{50}$s were determined with *Salmonella typhimurium* TA98 strain in the presence of S9 mix (containing 0.11 nmol P450 enzymes) as mentioned in the Text.

[c]The IC$_{50}$s for the N-hydroxylation of Trp-P-2 were determined with the microsomes from the AH22/pAMR2 cell (containing 0.052 nmol P450 1A1) and the excess amount of Trp-P-2 during the 3-min incubation as mentioned in the Text.

Antioxidative Potency and Desmutagenicity. On the hydroxyl positions, Terao et al. (*14*) and Cao et al. (*34*) described that the *ortho*-dihydroxyl structure in the B-ring (3' and 4') was necessary for the antioxidative activity of flavonoids. Cholbi et al. described free hydroxyl groups on the 5 and 7 positions in the A-ring, and/or on 3 position in the C-ring participated in the antioxidative effect (*32*). Luteolin has an *ortho*-dihydroxyl structure in the B-ring, but showed similar activity with other flavones not having the structure. Quercetin also has this structure, but was weaker than galangin. Among the flavonols, galangin to myricetin have 3-, 5-, and 7-hydroxyl groups. Their activity was not so different from the activity of flavonol and fisetin that lacked in any of three hydroxyls. For further examination, the IC$_{50}$s of hydroxy-masked flavonoids were determined. Isorhamnetin and hesperetin are masked one hydroxyl on the 3' or 4' position with methyl. Their IC$_{50}$s were lower than those of the original compounds, quercetin and eriodictyol. Even when all hydroxyl groups of quercetin and luteolin were blocked by methyl or acetyl, the IC$_{50}$s remained almost unchanged. These results indicated that the desmutagenicity of flavonoids related to the steric structure, but to neither hydroxyl number nor position.

Catechins have been found to be the strongest antioxidants among phytochemicals (*40-42*), and considered to contribute with the antioxidative potency associated with anticarcinogenicity (*43, 44*). Table III, however, shows that catechins failed the desmutagenicity against Trp-P-2. Catechin and epicatechin were non-effective, and epicatechin gallate was three-orders lower in the desmutagenicity than flavones and flavonols. Epigallocatechin also gave three-order lower IC$_{50}$ than quercetin and luteolin, as illustrated in Figure 1. Butylated hydroxyanisole and butylated hydroxytoluene are also strong phenolic antioxidants using popularly as food additives. Examining them (data not shown), the IC$_{50}$ of butylated hydroxyanisole was 400 nmol, and at more than 2500 nmol enhanced the mutagenicity of Trp-p-2. Butylated hydroxytoluene was non-effective between the concentration of 20 and 2600 nmol. The antioxidative potency was found to connect without the desmutagenicity.

P450 Inhibition and Desmutagenicity. To examine the function (b) in Table III, the IC$_{50}$s when converted into a molar ratio to P450 enzymes can give the better understanding, because the function (b) is an effect of flavonoids on the P450 enzymes. The S9 mix in the desmutagenicity test contained 0.11 nmol of P450 enzymes, and the inhibitory experiments used 0.052 nmol of P450 1A1. Thus, the desmutagenic IC$_{50}$s in the second column can compare with the doubled values of the inhibitory IC$_{50}$s in the third column. The desmutagenic IC$_{50}$s almost coincided with the respective IC$_{50}$s for the P450 inhibition. In both the tests, flavonoids from flavone to flavanone, and *all*-methoxyl luteolin and quercetin, gave the low values. Rutin and daidzein gave the larger values. Apigetrin showed the greatly large value for the desmutagenicity and then was non-effective on the P450 inhibition. Similarly, daidzein and puerarin were non-effective in both tests. The coincidences between both activities clearly indicate that the desmutagenicity of flavonoids was due to the function (b), this is, due to the inhibition of P450 1A1 that metabolically activates Trp-P-2 to the ultimate carcinogenic form.

An interesting point is the marked reduction in both the activities when flavonoids were transformed into 7- or 8-glycoside, but not 3-glycoside. The IC50 of apigenin was very low, 0.88 nmol for the desmutagenicity and 0.075 nmol for the P450 inhibition, but its 7-glycoside (apigetrin) gave 590 nmol for the desmutagenicity and became non-effective for the P450 inhibition. Similarly, daidzein led to fail the activities when derived to its 7-glycoside (daidzin) or to its 8-glycoside (puerarin). Contrary, the transformation of quercetin into the 3-glycosides (rutin and quercitrin) was shown to have IC50s not significantly different from the original quercetin. These marked changes in the IC50s were not detected in the smaller substituents, such as methyl and acetyl, as shown with the similar IC50s of all-methoxyl or acetyl quercetin and luteolin to those of the original compounds. Thus, the substitution of 7-or 8-hydroxyls with large group like glycoside greatly led to fail the activities, indicating that the steric structure around 7- and 8-carbons on flavonoid skeleton was important for the inhibition of P450 1A1.

High Affinity of Flavonoids to P450 1A1. Since flavones and flavonols among flavonoids tested here manifested the strong effect on the inhibition of N-hydroxylation by P450 1A1, the inhibitory types of flavones and flavonols were examined (Figure 2). When the activity of P450 1A1 constructed the Lineweaver-Burk plot, the curves in the presence of flavone, luteolin, chrysin, or apigenin crossed with the curve in the absence of them (control) on the x-axis. Thus, flavones were noncompetitive inhibitors. Flavonol, 3-hydroxy flavone, was also noncompetitive, but other flavonols showed the different types of the inhibition. Galangin and kaempferol gave the parallel curves to the control curve, indicating the uncompetitive type of curve. The curves of quercetin and myricetin crossed in the upper part of x-axis, indicating the mixed type. The inhibition types of flavonoids seemed to be classified according to the positions of hydroxyl groups. Flavones lacking in 3-hydroxyl and flavone having only 3-hydroxyl were noncompetitive inhibitors. Among flavonols, the additional two or three hydroxyl groups such as galangin and kaempferol acted uncompetitively, and more than four hydroxyls became the mixed type such as quercetin and myricetin.

When the Ki values of flavones and flavonols on the N-hydroxylation of Trp-P-2 by P450 1A1 were calculated with the results in Figure 2, they were in the range of 0.08 and 0.7 μM. Figure 3 shows that the Km value of Trp-P-2 as the substrate for P450 1A1 was 25 μM. Thus, the strong desmutagenicity of flavones and flavonols was contributed with the greatly lower Ki values on P450 1A1 comparing the Km of Trp-P-2.

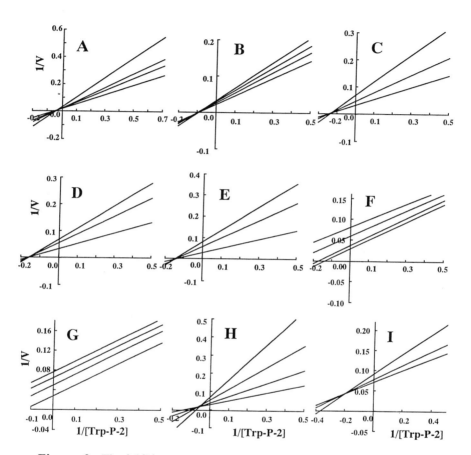

Figure 2. The inhibitory types of flavones and flavonols toward the N-hydroxylation of P450 1A1. When the N-hydroxylation of 1.5-7.8 μM Trp-P-2 (the concentration giving the linear increase in the reaction rate) was determined in the incubation system containing 0.025 nmol P450 1A1, flavones or flavonols were added at the concentrations near the *Ki* values. After three-separate experiments the reaction rates were plotted with the Lineweaver-Burk method. The alphabets in the figures show flavone (A), chrysin (B), apigenin (C), luteolin (D), flavonol (E), galangin (F), kaempferol (G), quercetin (H), and myricetin (I).

Discussion

We found that the antimutagenic mechanism of flavonoids against Trp-P-2 was due to the inhibition of P450 1A1, that metabolically activated Trp-P-2 to N-hydroxy-Trp-P-2, the ultimate carcinogenic form. The mechanism was the same, regardless of insolubilizing Trp-P-2 before being incorporated into the cells, neutralizing the N-hydroxy-Trp-P-2 (Table I), scavenging the radicals from N-hydroxy-Trp-P-2 before attack DNA (Table III), and suppressing the mutation fixation process after DNA has been damaged (Table II). The inhibitory effect on the P450 1A1 was remarkable in flavones and flavonols, and they had very low Ki values to the enzyme such as less than 1 μM comparing 25 μM for the Km value of Trp-P-2 (Figures 2 and 3). Flavones and flavonols with the high affinity showed the various types of inhibition on the N-hydroxylation by P450 1A1.

A region of the high affinity of flavones and flavonols to P450 1A1 seems to be constructed with 7- and 8-carbons, 1-pyran and B-ring (Figure 4). In Table III, apigenin and daidzein led to fail the activity when they were transformed to 7- or 8-glycoside, such as apigetrin, daidzin and puerarin. This marked reduction in the activity was not evoked by the 3-glycosidation as shown with rutin and quercitrin. Additionally, the smaller substitute groups such as methyl and acetyl did not induce the reduction. Thus, a large region around 7- or 8-carbon of flavonoids was important to interact with P450 1A1. The activity of isoflavones was weaker than that of flavones and flavonols. Since isoflavones bind the B-ring on 3-position, the B-ring also appeared to partly contribute to the affinity. Among flavanones that have the frequent C-ring, naringenin markedly reduced the activity when one hydroxyl group was added to 3' position in the B-ring as eriodictyol. Therefore, a region around 1-pyran encircled with the 7-carbon of A-ring and 3'-carbon of B-ring was considered to be the affinity site to P450 1A1.

The high affinity of flavones and flavonols to P450 1A1 implies that they were also able to be effective against other mutagens that activated by P450 1A1. Alldrick et al. (45) and Edenharder et al. (46) have reported that flavonoids suppressed the mutagenicity of other heterocyclic amines such as Trp-P-1, 2-amino-3-methylimidazo[4, 5-f]quinoline (IQ), 2-amino-3, 4-dimethylimidazo[4, 5-f]quinoline (MeIQ), 2-amino-3, 8-dimethylimidazo[4, 5-f]quinoxaline (MeIQx), and 2-amino-6-methyl-dipyrido[1, 2-a :3', 2'-d]imidazole (Glu-P-1). These heterocyclic amines also require the metabolic activation by P450 1A family (8, 30, 47). On the contrary, we have found that flavonoids were non-effective on the direct mutagens not requiring the activation, such as 1-nitropyrene (1-NP) and N-methyl-N'-nitro-N-nitrosoguanidine (MNNG) (2). Thus, the effect of flavones and flavonols are considered to be specific to the heterocyclic amines that are food-derived carcinogens.

Our daily meat includes the heterocyclic amines, and the vegetables provide flavonoids. The intake of vegetables at the same meal can mitigate the cancer risk. Flavones and flavonols with the strong desmutagenicity can prevent from the initiation step in carcinogenesis caused by the dietary carcinogens. Although the desmutagenicity of other flavonoids, isoflavones and catechins, were

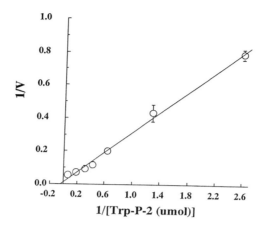

Figure 3. The *Km* value of Trp-P-2 to P450 1A1. Various amounts of Trp-P-2 as the substrate of P450 1A1 were added to the same incubation system as Figure 2, and then N-hydroxylation activity was determined. Constructing the Lineweaver-Burk plot after three-separate experiments, the *Km* value of Trp-P-2 to P450 1A1 was 25 μM.

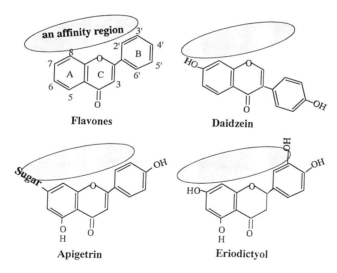

Figure 4. The proposed region for the affinity of flavones and flavonols to P450 1A1.

very weak, they have been well known to contribute with another function to the anticarcinogesis. The intake of isoflavones as daidzein and genistein including in a soybean closely correlate with a low mortality from prostatic cancer (*48*). The anticarcinogenicity has been found to be due to an induction of reversible G2/M arrest in cell cycle of the tumor cells (*15, 16*). The antioxidative potency of flavonoids can manifest the anticlastgenic effect (*49*), and should suppress the promotion step in carcinogenesis. This occurs, because active oxygen species participate in the promotion step (*13, 50*). Particularly, catechins have the strongest antioxidative potency (*40-42*). We conclude that flavones and flavonols contribute to the cancer prevention as the anti-initiator, isoflavones as the G2/M-arrest inducer to the tumor cells, and catechins as the anti-promoter.

Literature Cited

1. Natake, M.; Kanazawa, K.; Mizuno, M.; Ueno, N.; Kobayashi, T.; Danno, G.; Minamoto, S. *Agric. Biol. Chem.*. **1989**, 53, 1423-1425.
2. Kanazawa, K.; Kawasaki, H.; Samejima, K.; Ashida, H.; Danno, G. *J. Agric. Food Chem.* **1995**, 43, 404-409.
3. Samejima, K.; Kanazawa, K.; Ashida, H.; Danno, G. *J. Agric. Food Chem.* **1995**, 43, 410-414.
4. Hermann, K. *J. Food Technol.* **1976**, 11, 433-448.
5. Matsukura, N.; Kawachi, T.; Morino, K.; Ohgaki, H.; Sugimura, T. *Science* **1981**, 213, 346-347.
6. Yamaizumi, Z.; Shiomi, T.; Kasai, H.; Nishimura, S.; Takahashi, Y.; Nagao, M.; Sugimura, T. *Cancer Lett.* **1980**, 9, 75-83.
7. Knasmüller, S.; Kienzl, H.; Huber, W.; Herman, R.S. *Mutagenesis* **1992**, 7, 235-241.
8. Kato, R.; Kamataki, T.; Yamazoe, Y. *Environ. Health Perspect.* **1983**, 49, 21-25.
9. Wakabayashi, K.; Nagao, M.; Esumi, H.; Sugimura, T. *Cancer Res.* **1992**, 52, 2092s-2098s.
10. Sugimura, T. *Mutat. Res.* **1985**, 150, 33-41.
11. Nakayama, T. *Cancer Res.* **1994**, 54, 1991s-1993s.
12. Ubeda, A.; Esteve, M.L.; Alcaraz, M.J.; Cheeseman, K.H.; Slater, T.F. *Phytother. Res.* **1995**, 9, 416-420.
13. Rice-Evans, C.A.; Miller, N.J.; Paganga, G. *Free Radic. Biol. Med.* **1996**, 20, 933-956.
14. Terao, J; Piskula, M.K. in *Flavonoids in Health and Diseases*; Rice-Evans, C.A; Packer, L., Eds.; Marcel Dekker Inc.: NewYork, NY. 1997, pp. 277-293.
15. Matsukawa, Y.; Marui, N.; Sakai, T.; Satomi, Y.; Yoshida, M.; Matsumoto, K.; Nishino, H.; Aoike, A. *Cancer Res.* **1993**, 53, 1328-1331.
16. Plaumann, B.; Fritsche, M.; Rimpler, H.; Brandner, G.; Hess, R.D. *Oncogene* **1996**, 13, 1605-1614.

17. Tsyrlov, I.B.; Mikhailenko, V.M.; Gelboin, H.V. *Biochim. Biophys. Acta* **1994**, 1205, 325-335.
18. Siess, M.-H.; Leclerc, J.; Canivenc-Lavier, M.-C.; Rat, P.; Suschetet, M. *Toxicol. Appl. Pharmacol.* **1995**, 130, 73-78.
19. Obermeier, M.T.; White, R.E.; Yang, C.S. *Xenobiotica* **1995**, 25, 575-584.
20. Fuhr, U.; Kummert, A.L. *Clin. Pharmacol. Ther.* **1995**, 58, 365-373.
21. Mizuno, M.; Ohara, A.; Danno, G.; Kanazawa, K.; Natake, M. *Mutat. Res.* **1987**, 176, 179-184.
22. Sakaki, T.; Kominami, S.; Takemori, S.; Ohkawa, H.; Shibata, M. -A.; Yabusuki, Y. *Biochemistry* **1994**, 33, 4933-4939.
23. Sakaki, T.; Shibata, M.-A.; Yabusaki, Y.; Ohkawa, H. *J. Biol. Chem.* **1992**, 267, 16497-16502.
24. Bradford, M. *Anal. Biochem.* **1976**, 72, 248-254.
25. Minamoto, S.; Kanazawa, K. *Anal. Biochem.* **1995**, 225, 143-148.
26. Manabe, S.; Wada O. *Mutat. Res.* **1988**, 209, 33-38.
27. Yamazoe, Y.; Ishii, K.; Kamataki, T.; Kato, R.; Sugimura, T. *Chem. -Biol. Interact.* **1980**, 30, 125-138.
28. Ishii, K.; Ando, M.; Kamataki, T.; Kato, R.; Nagao, M. *Cancer Lett.* **1980**, 9, 271-276.
29. Wakata, A.; Oka, N.; Hiramoto, K.; Yoshioka, A.; Negishi, K.; Wataya, Y.; Hayatsu, H. *Cancer Res.* **1985**, 45, 5867-5871.
30. Kato, R.; Yamazoe, Y. *Jpn. J. Cancer Res.* **1987**, 78, 297-311.
31. Kada, T.; Shimoi, K. *BioEssay* **1987**, 7, 113-116.
32. Cholbi, M.R.; Paya, M.; Alcaraz, M.J. *Experientia* **1991**, 47, 195-199.
33. Jovanovic, S.V.; Steenken, S.; Tosic, M.; Marjanovic, B.; Simic, M.G. *J. Am. Chem. Soc.* **1994**, 116, 4846-4851.
34. Cao, G.; Sofic., E.; Prior, R.L. *Free Radic. Biol. Med.* **1997**, 22, 749-760.
35. Oeda, K.; Sakaki, T.; Ohkawa, H. *DNA* **1985**, 4, 203-210.
36. Sakaki, T.; Shibata, M.; Yabusaki, Y.; Ohkawa, H. *DNA* **1987**, 6, 31-39.
37. Arimoto, S.; Fukuda, S.; Itome, C.; Nakano, H.; Rai, H.; Hayatsu, H. *Mutat. Res.* **1993**, 287, 293-305.
38. Hayatsu, H.; Negishi, T.; Arimoto, S.; Hayatsu, T. *Mutat. Res.* **1993**, 290, 79-85.
39. Dashwood, R.; Yamane, S.; Larsen, R. *Environ. Mol. Mutagen.* **1996**, 27, 211-218.
40. Tournaire, C.; Croux, S.; Maurette, M.-T.; Beck, I.; Hocquaux, M.; Braun, A.M.; Oliveros, E. *J. Photochem. Photobiol. B: Biol.* **1993**, 19, 205-215.
41. Terao, J.; Piskula, M.; Yao, Q. *Arch. Biochem. Biophys.* **1994**, 308, 278-284.
42. Salah, N.; Miller, N.J.; Paganga, G.; Tijburg, L.; Bolwell, G.P.; Rice-Evance, C. *Arch. Biochem. Biophys.* **1995**, 322, 339-346.
43. Oguni, I.; Nasu, K.; Kanaya, S.; Ota, Y.; Yamamoto, S.; Nomura, T. *Jpn. J. Nutr.* **1989**, 47, 93-102.
44. Mukhtar, H.; Katiyar, S.K.; Agarwal, R. *J. Invest. Dermatol.* **1994**, 102, 3-7.
45. Alldrick, A.J.; Flynn, J.; Rowland, I.R. *Mutat. Res.* **1986**, 163, 225-232.

46. Edenharder, R.; Von Petersdorff, I.; Rauscher, R. *Mutat. Res.* **1993,** 287, 261-274.

47. Funae, Y.; Imaoka, S. in *Handbook of Experimental Pharmacology;* Schenkman, J.B.; Greim, H., Eds.; Springer-Verlag.: Berlin, Germany, 1993, Vol. 105; pp. 221-238.

48. Adlercreutz, H.; Markkanen, H.; Watanabe, S. *Lancet* **1993,** 342, 1209-1210.

49. Shimoi, K.; Masuda, S.; Furugori, M.; Esaki, S.; Kinae, N. *Carcinogenesis* **1994,** 15, 2669-2672.

50. Block, A. *Nutr. Rev.* **1992,** 50, 207-213.

Chapter 9

Cholesterol Lowering Effects of Red Koji

N. Nakajima[1] and A. Endo[1]

Department of Applied Biological Science, Tokyo Noko University, Fuchu, Tokyo 183, Japan

Red koji is a fermented food in East Asia, produced by growing fungi of the genus *Monascus*, over steamed rice. Several *Monascus* species are known to produce lovastatin, a 3-hydroxy-3-methylglutaryl coenzyme A (HMG-CoA) reductase inhibitor that lowers plasma cholesterol levels. The present study deals with the cholesterol lowering effects of a red koji (BY-114) in human subjects, with moderately elevated plasma cholesterol. Fifteen subjects were treated with a daily dose of one gram, for 6-8 weeks. In 6 of the subjects, the treatment did not bring an adequate response; reduction in plasma total, and LDL cholesterol was less than 10 percent. In the remaining 9 subjects, total and LDL cholesterol levels were reduced by over 10 percent. No significant adverse effects were observed throughout the trials.

Mevastatin (formerly called compactin), and lovastatin (formerly called mevinolin or monacolin K) (Figure 1), are specific inhibitors of 3-hydroxy-3-methyl coenzyme A (HMG-CoA) reductase, the rate-limiting enzyme in cholesterol biosynthesis (1). The former is produced by several species of the genus *Penicillium*, and the latter is a metabolite of *Aspergillus terreus*, and several species of the genus *Monascus*. Both mevastatin and lovastatin are extremely effective in lowering levels of plasma cholesterol in humans (1). The latter has been on the market as a cholesterol-lowering drug in the United States under the trade name Mevacor, since 1987.

Red koji, a fermented food in East Asia, is produced by growing several *Monascus* species over soaked and steamed rice, or koaliang. Red koji has been used for the production of wines and other fermented foods. It has also been an important folk medicine for over 600 years (2).

---

[1]Current address: Biopharm Research Laboratories, Inc., 2-1-31 Minamicho, Kokubunji, Tokyo 183, Japan.

In the present study, a type of red koji (BY-114) produced by a *Monascus* strain that produces small amounts of lovastatin, was tested for the ability to reduce levels of plasma cholesterol in healthy volunteers, with moderately elevated levels of plasma cholesterol. The results indicated that BY-114 was effective in lowering plasma cholesterol in most of these subjects.

Methods

Red koji (BY-114) was produced by growing *Monascus* sp. M1022 over steamed rice for several days (2). Fifteen clinically healthy volunteers with plasma total cholesterol levels of 240-290 mg/dl (10 men and 5 women), aged at 35-61 years, were enrolled for the present trials. All subjects gave informed consent at the onset of the trials. Subjects discontinued any lipid-lowering drugs at least 2 weeks before the start of the study. After the 2-week placebo period, subjects were treated with BY-114 for the next 6-8 weeks. BY-114 was given in one daily dose of one gram (dry weight) after their evening meal. After the withdrawal of BY-114, the subjects were given a placebo for 2 weeks. No instructions had been given to the subjects to refrain from eating or drinking.

Fasting blood samples were obtained periodically by venipuncture. Total cholesterol concentration was measured using a cholesterol esterase/cholesterol oxidase assay, according to the method of Allain et. al. (3). Plasma triglyceride was determined by enzymatically hydrolyzing triglycerides, and measuring released glycerol (4). Plasma high-density lipoprotein (HDL) cholesterol was determined after precipitation of apoprotein B, containing lipoproteins with magnesium phosphotungstate reagent (5). Plasma levels of low-density lipoprotein (LDL) were calculated from total cholesterol, HDL cholesterol, and triglycerides (6). Subjects were questioned about adverse experiences.

Results

The treatment at one gram did not bring an adequate response in the 6 subjects (cases 1 to 6), shown in Figure 2a. Thus, reduction of total cholesterol levels was less than 10 percent after a few weeks of treatment. In case 5, daily doses of BY-114 were raised to 2g after 4 weeks, resulting in a 14% reduction of total cholesterol at week 7, and thereafter. Six subjects, shown in Figure 2b (cases 7 to 12), gave an adequate response to BY-114. Total cholesterol levels were lowered to 220 mg/dl, or below, after 2 weeks of treatment in these subjects, except case 9, whose cholesterol was reduced to 240 mg/dl. The reduction of total cholesterol was 10 to 20 percent in these 6 cases; the decreases in LDL cholesterol were 13 to 25 percent. The lowered levels of total and LDL cholesterol were sustained throughout the trials. On the other hand, 3 subjects shown in figure 2c (cases 13 to 15) were highly sensitive to the treatment with BY-114: reduction of total cholesterol levels was 20 to 28 percent at week 2. In case 14, total cholesterol was reduced from 245 to 175 mg/dl at week 2. Administration of BY-114 to case 14 was discontinued, and resumed at week 4 at a daily dose of 0.5 g, giving a 14 percent decrease in total cholesterol. Some subjects reported a definite improvement in blood circulation, and reductions in constipation

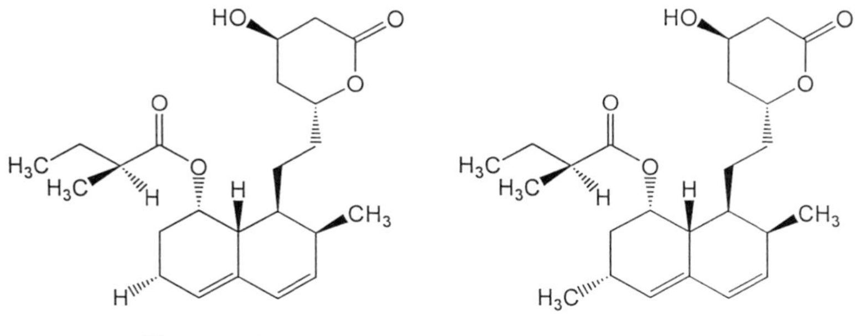

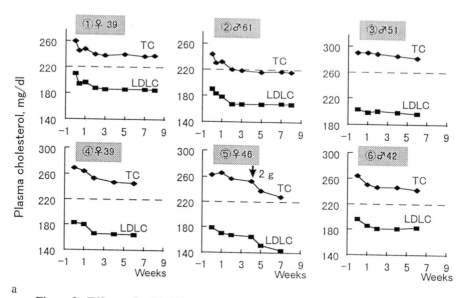

Figure 1. Mevastatin and lovastatin.

Figure 2. Effects of red koji BY-114 on levels of plasma total (TC) and LDL cholesterol (LDLC) in human subjects with mild hypercholesterolemia.

Continued on next page.

86

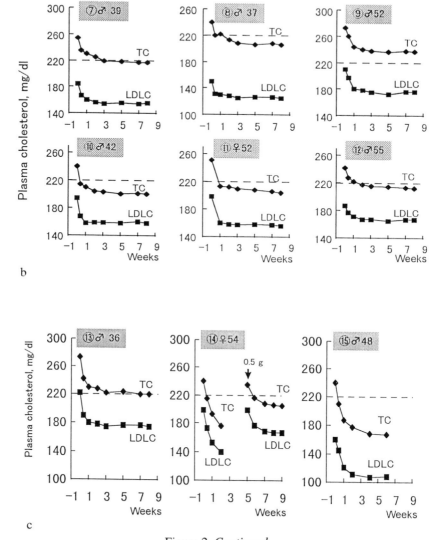

Figure 2. *Continued.*

and shoulder discomfort. In all subjects, the levels of LDL cholesterol were reduced in parallel with the reduction of total cholesterol levels. Those of HDL cholesterol were not changed detectably (data not shown). The reduced levels of both total, and LDL cholesterol came back to the pretreatment levels within 2 weeks after the termination of treatment in all subjects (data not shown).

No significant increases were observed in serum enzymes including GOT, GPT, and CPK during the treatment with BY-114. No adverse effects were reported by any of the subjects throughout the trials.

Discussion

The present trials have demonstrated that red koji (BY-114) reduced the levels of plasma total cholesterol by 10-28 percent, in 9 of the 15 mild patients with plasma total cholesterol levels of 240 to 290 mg/dl. According to our preliminary trials, BY-114 produced a 10-31 percent decrease in total cholesterol levels in 20 of the 31 patients with total cholesterol levels of 220 to 301 mg/dl, who had been treated at a daily dose of one gram, for four weeks. These findings indicate that BY-114 is effective in 2/3 of the patients with mild hypercholesterolemia.

Definite reductions in constipation and/or shoulder discomfort were reported by several subjects in both the present trials, and the preliminary experiments. It is likely that these pharmacological effects are caused by other constituents(s) of BY-114 than lovastatin, since it has not been reported that lovastatin and other HMG-CoA reductase inhibitors have such pharmacological effects.

In most hypercholesterolemic patients with a total cholesterol level below 260 mg/dl, lovastatin is effective in lowering plasma cholesterol at a daily dose of 2-5 mg (Endo, A., Tokyo Noko University, unpublished data). Each gram of red koji BY-114 contains around 1.5 mg of lovastatin. BY-114 also contains dietary fibre (3-5%), lipids (1-3%), proteins (7-9%), and total carbohydrates (70-80%). The cholesterol lowering effects of BY-114 may not be attributed solely to the presence of lovastatin. Rather, it is likely that dietary fibre, and other constituents found in BY-114, may work in concert with lovastatin to provide the cholesterol lowering effects.

Literature Cited

1. Endo, A. *J. Lipid Res.* **1992**, 33, 1569-1582.
2. Endo, A. *Fermentation and Industry*. **1985**, 43, 544-522.
3. Allain, C. C.; Poon, L. S., et al. *Clin. Chem.* **1974**, 20, 470-475.
4. Bucolo, G.; David, H. *Clin. Chem.* **1973**, 19, 476-482.
5. Finley, P.R.; Schifman, P.B., Williams, R.T. et al. *Clin. Chem.* **1978**, 24, 931-933.
6. Friedewalde, W.T.; Levy, R.I., Fredrickson, D.S. *Clin. Chem.* **1972**, 18, 499-502.

Chapter 10

Heterocyclic Amines Induce Apoptosis in Hepatocytes

Hitoshi Ashida, Hideya Adachi, Kazuki Kanazawa, and Gen-ichi Danno

Department of Biofunctional Chemistry, Faculty of Agriculture, Kobe University, Kobe 657, Japan

Heterocyclic amines in food are well known as mutagens and carcinogens. We have examined the cytotoxicity of heterocyclic amines to the primary cultured hepatocytes of rats. Tryptophan-pyrolysate, 3-amino-1,4-dimethyl-5H-pyrido[4,3-b]indole (Trp-P-1), was the most toxic compound among seven heterocyclic amines, and four structural analogues for Trp-P-1. Trp-P-1 caused typical apoptotic morphology in cell and nuclear structures, and induced DNA fragmentation in a time- and dose-dependent manner. Evidence for the activation of endonuclease was obtained by using zinc ion as an inhibitor. In addition, Trp-P-1 induced p53 and c-myc proteins. This data indicates that Trp-P-1 induces apoptotic cell death to hepatocytes. We suggest that liver cells, when exposed to carcinogens, have a latent ability to avoid malignant alterations by inducing apoptosis and killing themselves.

Charcoal broiled food contains various kinds of carcinogenic heterocyclic amines and is a particularly potent bacterial mutagen. Among the heterocyclic amines, tryptophan pyrolysis products, Trp-P-1 (3-amino-1,4-dimethyl-5H-pyrido[4,3-b]indole) and Trp-P-2 (3-amino-1-methyl-5H-pyrido[4,3-b]indole) showed strong mutagenicity by a bacterial test using *Salmonella* strain (*1*). However, Trp-P-1 and Trp-P-2, themselves, do not have mutagenic activity. They are promutagens which can be converted to their *N*-hydroxy forms by cytochrome P-450 enzymes mainly in liver, and subsequently show the mutagenic and carcinogenic activities (*2*). Activated Trp-P-1 and Trp-P-2 affect DNA damage and cell viability in the bacterial strains. Previously, we found that kaempferol coumarate in bay, galangin in oregano, and luteolin in sage and thyme were potent desmutagens against Trp-P-2 in bacterial test

(*3, 4*). However, few papers have been published concerning the toxicity of heterocyclic amines such as Trp-P-1 and Trp-P-2 to mammalian cells. It is, therefore, interesting to study the cytotoxicity and toxic mechanisms of heterocyclic amines in primary cultured hepatocytes of rats.

In this study, the authors have investigated the cytotoxicity of heterocyclic amines, and their related compounds to primary cultured hepatocytes of rats. It has been found that Trp-P-1 is the most toxic compound among seven heterocyclic amines and four analogue compounds. We also have observed several hallmarks of apoptosis, including DNA ladder, chromatin condensation and fragmentation, and an induction of apoptosis-associated proteins, p53 and c-myc, after treatment with Trp-P-1. Finally, we will discuss the relationship between the mutagenicity and apoptosis, both induced by Trp-P-1. This study is significant to recognize the protective mechanism from carcinogens in a living body.

Primary culture of hepatocytes and treatments.

Six-to-ten-week old male Wistar rats (Japan SLC, Inc., Shizuoka, Japan) were housed in suspended steel cages, provided with commercial chow and water *ad libitum*, and maintained in a controlled room (temperature 25°C, 12 h light-dark cycle). Parenchymal hepatocytes were isolated from rats by *in situ* perfusion of liver with collagenase solution by the method of Tanaka *et al* (*5*). The isolated hepatocytes were suspended at a concentration of 5×10^5 cells/ml into William's medium E, with 100 nM insulin, 100 nM dexamethasone, 100 mg/l kanamycin, 10 KIU/ml aprotinin and 5% fetal bovine serum. The cells were seeded on the plastic multiplates or dishes (Becton Dickinson Co., Ltd., Flanklin lakes, NJ) precoated with collagen, and cultured in an atmosphere of 95% air-5% CO_2 at 37°C for 2 h. Primary cultured hepatocytes were treated with a various concentration of heterocyclic amines and their analogues in dimethyl sulfoxide (DMSO) for definite time points, as indicated in each figure or table. Structures of heterocyclic amines and their analogues are shown in Figure 1. Control cells were treated with the same volume of vehicle (final 0.1% DMSO). These cells were submitted to the following experiments.

Cytotoxicity of Heterocyclic Amines to Primary Cultured Hepatocytes

Cytotoxicity of seven heterocyclic amines and four analogues of tryptophan pyrolysates, to primary cultured hepatocytes were estimated by the measurement of cell viability and the lactate dehydrogenase (LDH) leakage test. The cell viability was measured by the MTT test (*6, 7*) after the treatment with each chemical at 60 μM for 6, 12 and 24 h. As shown in Table I, Trp-P-1 strongly suppressed the cell viability of hepatocytes to 20%. Trp-P-2 also suppressed it significantly to 50% 6 h after treatment. Other heterocyclic amines, PhIP, IQ, MeIQ, MeIQx and A[α]C did not show any cytotoxicity to hepatocytes by 24 h. Harmine, harmane, norharmane, and harmol slightly reduced the cell viability, but their toxic action was weaker than that of Trp-P-1 or Trp-P-2.

90

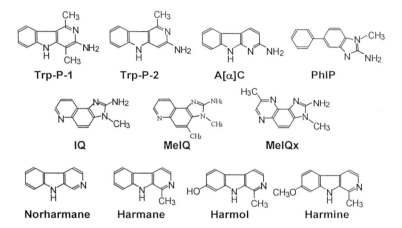

Figure 1. Structure of heterocyclic amines and their analogue compounds.

For the LDH leakage test, after the hepatocytes were treated with each chemical, culture medium was stored, and the cells were homogenized with 0.05 M phosphate buffer (pH 7.4) by ultrasonication. LDH activity was measured both in culture medium, and in cell homogenates, according to the method of Bergmeyer et al. (8). LDH leakage is represented as the percent of enzyme activity in medium against the sum of medium and intracellular activities. Trp-P-1 and Trp-P-2 caused significant LDH leakage 6 h after treatment. The leakage of the cells treated with Trp-P-1 at 30 and 60 μM was 75 ± 12 and $95 \pm 8\%$, respectively, and that of the control cells was $3.8 \pm 2.4\%$. Trp-P-2 at 120 and 240 μM also increased in the leakage to 19 ± 5 and $98 \pm 11\%$, respectively. Other chemicals did not show any significant leakage. This data indicates that tryptophan pyrolysates cause the cytotoxicity to hepatocytes, and the toxicity of Trp-P-1 is stronger than that of Trp-P-2.

Table I. Effect of heterocyclic amines and their analogues on cell viability in primary cultured hepatocytes of rats

Chemicals	Treatment time (h)		
	6	12	24
	Cell viability (% of control)		
Control	100 ± 9	100 ± 12	100 ± 14
Trp-P-1	$19 \pm 2*$	$9 \pm 1*$	$7 \pm 2*$
Trp-P-2	$48 \pm 6*$	$31 \pm 4*$	$9 \pm 1*$
PhIP	109 ± 10	101 ± 16	110 ± 10
IQ	101 ± 7	104 ± 15	112 ± 9
MeIQ	106 ± 6	101 ± 15	98 ± 13
MeIQx	110 ± 12	94 ± 9	85 ± 8
A[α]C	108 ± 9	111 ± 12	104 ± 9
Harmine	83 ± 10	84 ± 8	80 ± 14
Harmane	85 ± 12	85 ± 16	81 ± 20
Norharmane	81 ± 11	77 ± 12	75 ± 15
Harmol	82 ± 9	$69 \pm 9*$	$65 \pm 5*$

Hepatocytes were treated with each chemical at 60 μM for 6 h. The cell viability was measured by the MTT test. Data are expressed as mean $\pm$ SD (n=3). Asterisks show significant difference from corresponding control value (p<0.05).

Among eleven compounds tested here, Trp-P-1 drastically lost the cell viability and Trp-P-2 showed moderate action. LDH leakage showed the same tendency on the cell viability. It is difficult to elucidate the relationship between the toxic activity and chemical structures of heterocyclic amine type chemicals. The toxic activity of these compounds was as follows; Trp-P-1□ Trp-P-

2□ harmol□ others. At least the imidazoquinoline or imidazopyridine ring did not contribute to the toxicity, because IQ, MeIQ, and PhIP did not show the toxicity to hepatocytes. A prydoindole ring with an amino group was necessary for the cytotoxicity. Among Trp-P-1, Trp-P-2 and A[α]C, the toxicity was dependent upon a number of methyl groups. Interestingly, the cytotoxic activity did not correlate with the mutagenicity, because the mutagenicity of Trp-P-1 is weaker than that of Trp-P-2, and harmine, harmane, norharmane, and harmol are co-mutagens, but not mutagens. Further investigation will be needed to obtain information about the structure-activity relationship.

Heterocyclic Amines Induce DNA-fragmentation to Hepatocytes

Since loss of cell viability is closely relate to apoptosis, it was investigated whether heterocyclic amines and their analogues can induce apoptosis. DNA fragmentation was measured, which is a biochemical marker for apoptosis. Briefly, hepatocytes (1×10^6 cells on 35 mm-dish) were treated with each chemical for 6 h as described above and harvested with 200 µl TE buffer (10 mM Tris-HCl, pH 7.4, 10 mM EDTA) containing 0.5% Triton X-100. The cells were stored on ice for 10 min at 4° C to allow the lysis of the plasma membrane. The lysate was centrifuged at 400 x g for 5 min and separated into the supernatant containing fragmented DNA, and precipitated nucleus containing intact DNA. The precipitated nucleus was lyzed with 200 µl TE buffer containing 0.5% SDS. To remove RNA and protein, both intact and fragmented DNA fractions were incubated with 500 µg/ml RNase A at 50°C for 30 min and, subsequently, with 500 µg/ml proteinase K at 50°C for 60 min. After the addition of 0.5 M NaCl and 1 mM EGTA at the final concentrations, DNA was precipitated in 50% isopropanol at -20°C overnight. DNA was collected by a centrifugation at 15,000 rpm for 20 min and washed with 70% ethanol. Both fragmented and intact DNA was re-suspended into 500 µl TE-Na buffer (50 mM Tris-HCl, pH 7.5, 5.0 mM EDTA and 0.1 M NaCl). DNA content was measured in both fractions using 4',6-diamidino-2-phenylindole dihydrochloride (DAPI), as a specific fluorescence probe, for double-stranded DNA (9). The percent of DNA fragmentation was calculated, and data was expressed as a ratio to control. The results are shown in Figure 2. Trp-P-1 increased in DNA fragmentation in a dose-dependent manner. DNA fragmentation of the cells treated with 30 µM Trp-P-1 was increased to 14-fold as much as that of control cells, and was down-regulated at 60 µM. Trp-P-2 and harmol (60 µM each) also caused DNA fragmentation, but other chemicals did not affect DNA fragmentation.

DNA fragmentation, that is chromatin fragmentation into oligonucleosomes, was also analyzed by an agarose gel electrophoresis. The fragmented DNA fraction was prepared from a different culture series. DNA was placed onto a 2% agarose gel, and electrophoresis was performed at 100 V for 1 h. After the electrophoresis, the gel was stained with ethidium bromide, and fragmented DNA was analyzed by transilluminator. As shown in Figure 3A, Trp-P-1 caused DNA ladder more than 10 µM 6 h after treatment. When hepatocytes were treated with 30 µM Trp-P-1, we detected a

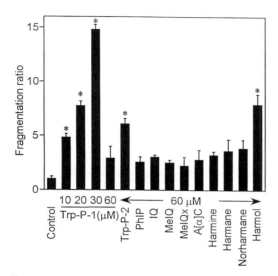

Figure 2. Effect of heterocyclic amines and their analogues on DNA fragmentation in hepatocytes. Hepatocytes were treated with 60 µM of test compounds or DMSO as a vehicle for 6 h. DNA fragmentation was quantified by using a fluorescence probe, DAPI.

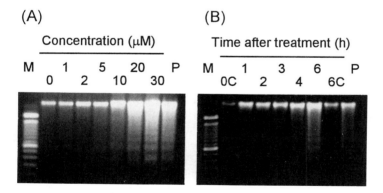

Figure 3. Time-course and dose-response effect of Trp-P-1 on DNA fragmentation in hepatocytes. Hepatocytes were treated with 30 µM Trp-P-1 for indicated time (panel A) or with various concentrations of Trp-P-1 for 6 h (panel B). DNA fragmentation was analyzed by a 2% agarose gel electrophoresis. M, 100 bp marker; C, control; P, positive control (1 µg/ml actinomycin D and 10 ng/ml TNF α).

DNA ladder pattern 4 h after treatment (Figure 3B). These results indicate that Trp-P-1 induces DNA fragmentation with a time- and dose-dependent manner. It is well known that DNA fragmentation is catalyzed by endonucleases in apoptotic cells. To confirm an activation of endonucleases by Trp-P-1, the effect of Zn^{2+} as the endonuclease inhibitor, on Trp-P-1-induced DNA fragmentation, was examined. Ratio of Trp-P-1-induced DNA fragmentation (14.2 ± 1.5) was suppressed by a pretreatment with 1 mM Zn^{2+} (3.6 ± 1.1). These data indicate Trp-P-1 activates a certain zinc-sensitive endonucleases such as endonuclease γ (DNase γ) and NUC18.

Analysis of Cell Morphology

Primary cultured hepatocytes were cultured on a plastic cover slip, and treated with 30 μM Trp-P-1 for 6 h. The cells were washed with PBS twice, fixed with 1% glutaraldehyde in PBS for 30 min and analyzed the cell morphology with a phase contrast microscope. As shown in Figure 4, the structure of Trp-P-1-treated hepatocytes was disintegrated into large blebs with diameters of 5 to 10 μm. We also analyzed nucleus structure using a fluorescence microscope after staining the cells with 1.5 mg/ml Hoechst 33258 for 30 min. Typical chromatin condensations were observed in Trp-P-1-treated hepatocytes (data not shown). Neither cell, nor nuclear morphology changes were observed in the control cells. From these morphological changes, and a significant increase in DNA fragmentation, we confirm that Trp-P-1 induces apoptosis in the liver cells.

Induction of c-Myc and p53 Proteins in Trp-P-1-treated Hepatocytes

Nuclear transcription factors such as p53 and c-myc, are intimately associated with cellular proliferation and are also important in apoptosis (*10-13*). Nuclear protein extracts were prepared from the cells treated with 30 μM Trp-P-1 for 15 to 360 min. Briefly, cultured hepatocytes on a 100 mm-dish were harvested and homogenized with a 400 μl of homogenizing buffer (10 mM Tris-HCl, pH 7.9, 1.5 mM $MgCl_2$, 10 mM KCl, 0.5 mM dithiothreitol (DTT), 0.5 KIU/ml aprotinin, and 1.0 μg/ml leupeptin). The homogenate was centrifuged at 400 x *g* for 5 min at 4°C to pellet crude nuclei. The pellet was washed twice with the homogenizing buffer containing 0.5% Triton X-100, and once with the homogenizing buffer under the same centrifugation condition. The pellet was, then, homogenized with 30 μl extraction buffer (10 mM Tris-HCl, pH 7.9, 25% glycerol, 0.42 M NaCl, 1.5 mM $MgCl_2$, 0.2 mM EDTA, 1.0 mM phenylmethylsulfonylfloride (PMSF), 0.5 mM DTT, 0.5 KIU/ml aprotinin, and 1.0 μg/ml leupeptin). The homogenate was kept on ice for 30 min with an occasional mixing, and was centrifuged at 16,000 x *g* for 20 min at 4°C. Clear supernatant was collected, and used as a nuclear protein fraction. The protein concentration in the nuclear protein fraction was determined by the method of Lowry *et al.* (*14*), using bovine serum albumin as a standard protein.

A Western blotting analysis with corresponding monoclonal antibodies against p53 and c-myc detected the expression of p53 and c-myc proteins in the nuclear protein fraction. The nuclear protein fraction (30 μg protein each) was subjected to a 10% SDS-poly acrylamide gel electrophoresis, and transferred to a poly(vinylidene difluoride) (PVDF) filter. After blocking the PVDF filter with 5% non-fatted dried milk in TBST buffer (10 mM Tris-HCl, pH 8.0, 150 mM NaCl and 0.06% Tween 20) at 4°C overnight, the filter was incubated with the primary antibodies against c-myc or p53 at room temperature for 1 h. The filter was washed with TBST buffer several times, and was incubated with horseradish peroxidase-linked secondary antibodies at room temperature for 30 min. The filter was washed again with TBST buffer several times, immunocomplexes were visualized with the ECL Western blotting chemiluminescence-detecting reagent (Amersham).

The induction of p53 was detected 15 min after treatment with Trp-P-1, as shown in Figure 5A. The levels of p53 increased time-dependently by 2 h, and decreased 3 h after treatment. After 6 h-treatment, p53 was not detectable. Interestingly, at least two junk bands (30 and 35 kDa) were observed by the Trp-P-1-treatment. Since the monoclonal antibody used here can detect both wild and denatured (or mutant) p53 proteins, these protein bands may be denatured p53 protein. As shown in Figure 5B, c-myc protein was also detected 15 min after the treatment and increased in time-dependently. Thus, Trp-P-1 caused a transient up-regulation, a subsequent down-regulation of p53 protein, and an apparent up-regulation of c-myc protein. This data clearly indicates that Trp-P-1 stimulates nuclear transcription factors accompaned with the induction of cell death signal.

Tumor suppresser gene product, p53, exerts a significant and dose-dependent effect in initiation of apoptosis, but only when it is induced by agents that cause DNA-strand breakage (11). On the other hand, transcription factor, c-myc, is one of the proto-oncogene products, and is intimately associated with cellular prolif-eration in the presence of appropriate growth factors. Expression of c-myc works as a potent inducer of apoptosis when cell proliferation is blocked by deprivation of growth factors, or forcible arrest with cytostatic drugs (10, 15). In an apoptotic pathway, c-myc requires the presence of a functional p53 protein, thus, p53 may mediate apoptosis as a safeguard mechanism to prevent cell proliferation induced by oncogene activation (12). The induction of both p53 and c-myc, by Trp-P-1, accompanied with DNA fragmentation, is apparently identical with the apoptotic mechanism described above.

Relationship between the mutagenicity and apoptosis

Trp-P-1 and Trp-P-2 are known to be metabolized to their N-hydroxy forms by cytochrome P450s, probably by P448 (16). N-Hydroxy heterocyclic amine generates superoxide anion (17) and subsequently forms DNA adduct (18, 19). To examine whether the Trp-P-1-induced cytotoxicity was due to the reactive oxygen species-mediated lipid peroxidation, thiobarbituric acid reactive substances (TBARS) were evaluated in hepatocytes treated with Trp-P-1 or Trp-P-2 , for 1 h, by the method of Ohkawa et al.(20). Both Trp-P-1 and Trp-P-2 did not increase in TBARS level in primary cultured hepatocytes (Table II). Thus, the toxicity of Trp-P-1 and Trp-P-2 on cells would not be due to the lipid peroxidation. To know whether cultured hepatocytes of untreated rats have an ability to metabolize Trp-P-1 to its N-hydroxy

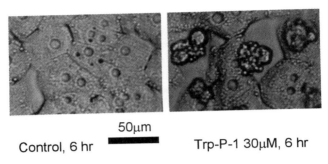

50μm

Control, 6 hr   Trp-P-1 30μM, 6 hr

Figure 4. Cell morphology of Trp-P-1-treated hepatocytes with a phase contrast microscope.

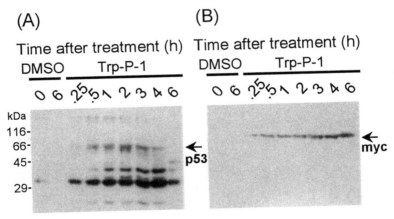

Figure 5. Western blotting analysis of p53 (A) and c-myc (B) in Trp-P-1-treated nuclear protein. Nuclear protein extracts (30 μg) from 30 μM Trp-P-1-treated hepatocytes were subjected to a 10% SDS-PAGE and transferred to a PVDF membrane. The immunoblot analysis was carried out with a monoclonal antibody against p53 or c-myc.

form, we also examined the loss of the P450 content in hepatocytes during the culturing process. Hepatocytes were isolated from untreated rats and cultured for 2 and 8 h. The cells were washed and homogenized with PBS. Homogenates were centrifuged at 9,000 x g for 20 min, and the supernatant was used as an S-9 fraction. The total cytochrome P450 content in the S-9 fraction was measured according to the method of Omura and Sato (21). As shown in Figure 6, the P450 content declined during the culture of hepatocytes, and it almost completely disappeared by 8 h. The loss of P450 content relates to the loss of the activation of Trp-P-1, to its N-hydroxy form. We attempted to measure the mutagenicity against 100 ng Trp-P-1, or 20 ng Trp-P-2, with this S-9 fraction by Ames test with Salmonella typhimurium TA98 (Table III). Trp-P-1 and Trp-P-2 incubated in the presence of the S-9 fraction from untreated hepatocytes (2 or

Table II. Effect of Trp-P-1 and Trp-P-2 on lipid peroxidation in primary cultured hepatocytes of rats

Treatment	Concentration (μM)	TBARS (nmol/mg protein)
DMSO	-	2.58 $\pm$ 0.20
Trp-P-1	10	1.96 $\pm$ 0.31
	20	2.79 $\pm$ 0.21
	30	2.03 $\pm$ 0.10
	60	1.86 $\pm$ 0.11
Trp-P-2	60	2.15 $\pm$ 0.11

Hepatocytes were treated with Trp-P-1 and Trp-P-2 for 1 h, and the amount of TBARS was measured. Data are expressed as mean $\pm$ SD (n=3).

Table III. Loss of the mutagenic activity in hepatocytes during culture

Time in culture (h)	Revertants/plate (TA 98)[a]	
	Trp-P-1 (100 ng/plate)	Trp-P-2 (20 ng/plate)
t0 (isolated hepatocytes)	142 $\pm$ 13	210 $\pm$ 13
t2 (2 h cultured hepatocytes)	94 $\pm$ 29	18 $\pm$ 2
t8 (8 h cultured hepatocytes)	20 $\pm$ 15	23 $\pm$ 8
MC + PB[b]	2710 $\pm$ 51	991 $\pm$ 68

The mutagenicity was measured by the Ames test with Salmonella typhimurium TA98 strain and the S-9 fraction (0.40 mg protein/plate) from untreated rat hepatocytes. a, Number of revertants were subtracted spontaneous revertants (17 $\pm$ 2). b, The S-9 fraction from methylcholanthrene and phenobarbital pretreated rat liver. Data are expressed as mean $\pm$ SD (n=3).

98

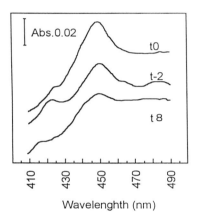

Wavelenghth (nm)

Figure 6. Loss of the reduced carbon monoxide spectrum of P450 in hepatocytes during culture. The S-9 fraction was prepared from untreated rat hepatocytes and the total cytochrome P450 content were measured. t0 is the spectrum for freshly isolated hepatocytes, t2 and t8 are spectra for 2 and 8 h cultured hepatocytes, respectively. The microsomal protein concentration for each determination was 4.47 mg/ml.

8 h in culture) and bacteria gave rise to the small number of revertants, which doubled the level of spontaneous mutants. The number of the revertants declined during the culturing of hepatocytes, with the same manner of the P450 content. When we used the S-9 fraction of liver prepared from methylcholanthrene and phenobarbital pretreated rats, 100 ng Trp-P-1 and 20 ng Trp-P-2 gave about 2,700 and 990 revertants, respectively. These results indicate that primary cultured hepatocytes of the untreated rat lost the metabolic activity of Trp-P-1 and Trp-P-2 to N-hydroxy forms. Similar results were also reported using a *Salmonella*-hepatocytes system (*22*). Thus, we suppose that the mechanism of Trp-P-1-induced apoptosis is different from that of the mutagenicity and it may not require a drug-metabolizing system. This is perhaps not surprising since Trp-P-1 intercalates DNA and acts as a DNA repair-inhibitor without metabolic activation (*19*). It is, however, difficult to prove the relationship between the mutagenicity and apoptosis, both induced by Trp-P-1 from these data.

Conclusions

In this study, we have compared the cytotoxicity of heterocyclic amines to that of primary cultured hepatocytes of rats. Among seven heterocyclic amines and four analogues, Trp-P-1 showed the strongest cytotoxicity to hepatocytes. Trp-P-1 caused DNA fragmentation in a time- and dose-dependent manner. Trp-P-1 induced apoptosis related proteins, p53 and c-myc, in nuclear extracts. This data indicates that Trp-P-1 induces apoptotic cell death in hepatocytes. We suppose that the induction of apoptosis is one of the defense systems due to liver specific carcinogens, such as Trp-P-1 and Trp-P-2. Normal cells, when exposed to carcinogens, have a latent ability to avoid malignant alterations by inducing apoptosis and killing themselves. The relationship between apoptosis and the mutagenicity, both caused by heterocyclic amines, is still not clear and will be investigated in the future.

Literature Cited

1. Negishi, T; Hayatsu, H. *Biochem. Biophys. Res. Commun.* **1979**, *88*, 97-102.
2. Kato, R; Yamazoe, Y. *Jpn. J. Cancer Res.* **1987**, *78*, 297-311.
3. Kanazawa, K.; Kawasaki, H.; Samejima, K.; Ashida, H.; Danno, G. *J. Agric. Food Chem.* **1995**, *43*, 404-409.
4. Samejima, K.; Kanazawa, K.; Ashida, H.; Danno, G. *J. Agric. Food Chem.* **1995**, *43*, 410-414.
5. Tanaka, K.; Sato, M.; Tomita, Y.; Ichihara, A. *J. Biochem.*, **1978**, *84*, 937-946.
6. Mosmann, T. *J. Immunol. Methods* **1983**, *65*, 55-63.
7. Oka, M., Maeda, S., Koga, N., Kato, K., and Saito, T. *Biosci. Biotech. Biochem.* **1992**, *56*, 1472-1473.
8. Bergmeyer, H. U.; Graßl, M.; Walter, H. -E. In *Methods of enzymatic analysis*, 3 rd ed. Vol. 2; Bergmeyer, H. U., Bergmeyer, J., and Graßl, M. Eds.; Verlag Chemie, Weinheim 1983; pp. 126-328.

9. Brunk, C. F.; Jones, K. C.; James, T. W. *Anal. Biochem.* **1979**, *92*, 497-500.

10. Evan, G. I.; Wyllie, A. H.; Gilbert, C. S.; Littlewood, T. D.; Land, H.; Brooks, M.; Waters, C. M.; Penn, L. Z.; Hancock, D. C. *Cell*, **1992**, *69*, 119-128.

11. Clarke, A. R.; Purdie, C. A.; Harrison, D. J.; Morris, R. G.; Bird, C. C.; Hooper, M. L.; Wyllie, A. H. *Nature* **1993**, *362*, 849-852.

12. Hermeking, H.; Eick, D. *Science* **1994**, *265*, 2091-2093.

13. Jiang, M. -C.; Yang-Yen, H. -F.; Lin, Jen. -Kun.; Yen, J. J. -Y. *Oncogene* **1996**, *13*, 609-616.

14. Lowry, O. H.; Rosebrough, N. J.; Farr, A. L.; Randall, R. J. *J. Biol. Chem.* **1951**, *193*, 265-275.

15. Evan, G.; Harrington, E.; Fanidi, A.; Land, H.; Amati, B.; Bennett, M. *Philos. Trans. R. Soc. Lond. B. Biol. Sci.* **1994**, *345*, 269-275.

16. Yamazoe, Y.; Ishii, K.; Kamataki, T.; Kato, R.; Sugimura, T. *Chem. -Biol. Interact.* **1980**, *30*, 125-138.

17. Wakata, Y.; Yamane, K.; Hiramoto, K.; Ohtsuka, Y.; Okubo, Y.; Negishi, K.; Hayatsu, H. *Jpn. J. Cancer Res.* **1988**, *79*, 576-579.

18. Kato, R.; Yamazoe, Y. *Jpn. J. Cancer Res.* **1987**, *78*, 297-311.

19. Shimoi, K.; Miyamura, R.; Mori, T.; Todo, T.; Ohtsuka, E.; Wakabayashi, K.;Kinae, N. *Carcinogenesis* **1996**, *17*, 1279-1283.

20. Ohkawa, H.; Ohishi, N.; Yagi, K.; *Anal. Biochem.*, **1979**, *95*, 351-358.

21. Omura, T.; Sato, R. *J. Biol. Chem.* **1964**, *239*, 2370-2378.

22. Decloître, F.; Hamon, G.; Martin, M.; Thybaud-Lambay, V. *Mutat. Res.* **1984**, *137*, 123-132.

Chapter 11

Antioxidative Activity of Brewed Coffee Extracts

Apasrin Singhara, Carlos Macku[1], and Takayuki Shibamoto[2]

Department of Environmental Toxicology, University of California,
Davis, CA 95616

Dichloromethane extracts and their components isolated from brewed coffee were evaluated for antioxidative activity, measured by the oxidative converson of pentanal or hexanal to a corresponding acid. Extracts were fractionated using column chromatography and each fraction was tested for antioxidative activity. A fraction eluted from the headspace sample with a pentane/ethyl acetate (95/5) solution inhibited the pentane/pentanoic acid conversion for more than 10 days. A fraction eluted from a dichloromethane extract of brewed coffee with 100% acetone inhibited the hexanal/hexanoic acid conversion by 100% for more than 14 days. Maltol and 5-hydroxy methylfurfural (5-HMH) were identified in the fraction eluded with 100% acetone as major components. Maltol inhibited the acid formation by 100% at levels higher than 250 mg/mL. Dose-related inhibitory activity was observed in the case of 5-HMF which inhibited the acid formation by 95% and 50% at the levels of 500 mg/mL and 50 mg/mL, respectively. These activities were comparable to those of known antioxidants, BHT and a-tocopherol in this testing system.

Production of antioxidative compounds in processed foods has been reported many times in literature. Formation of these antioxidants is most likely related to the Maillard reaction. In the late 1950s and early 1960s, the addition of sugar and/or amino acids to baked foods, such as cookies, was found to enhance the browning reaction that subsequently increased their stability against oxidative rancidity (1, 2). It was obvious that the baking process produced some antioxidants. Many researchers reported that heat treatment improved the oxidative stability of various foods, including milk products (3), cereals (4), and meats (5).

The higher molecular weight substances, such as melanoidins, produced from a sugar/amino acid model system by the Maillard reaction, have received much attention as antioxidants. For example, melanoidin and its ozone oxidation products

---

[1]Current address: Planters LifeSavers Company, 200 DeForest Avenue, East Hanover, NJ 07936.
[2]Corresponding author.

significantly inhibited linoleic acid oxidation (*6*). Recently, low molecular weight volatile compounds obtained from a glucose/cysteine browning model system were reported to possess certain antioxidative activities (*7*). Also, column chromatographic fractions prepared from a dichloromethane extract of a glucose/cysteine browning model system inhibited the oxidative transformation from hexanal to hexanoic acid. Additionally, several nitrogen- and/or sulfur-containing heterocyclic compounds, which are major flavor compounds formed by the Maillard reaction (*8*), exhibited antioxidative activity in two testing systems (*9*).

In the present study, antioxidative activity of volatile compounds obtained from a dichloromethane extract of brewed coffee, which reportedly contained numerous heterocyclic flavor compounds (*10*), was investigated.

Experimental

Materials. Pentanal, hexanal, 2,5-dimethylhexane, 5-hydroxymethylfurfural (5-HMF), maltol, caffeine, and a-tocopherol were purchased from Aldrich Chemical Co. (Milwaukee, WI). Butylated hydroxymethyl toluene (BHT) was bought from Sigma Chemical Co. (St. Louis, MO). All authentic samples were obtained from reliable commercial sources.

Regular and decaffeinated ground roasted coffees were purchased from a local market.

Sample Preparation with Simultaneous Purging and Solvent Extraction (SPE). Ground roasted regular coffee (25 g) was brewed using a stove-top coffee brewer with 300 mL tap water. The freshly brewed coffee was mixed with 17.5 g of sodium chloride. The solution was placed in a 500 mL, two-neck, round-bottom flask interfaced to a simultaneous purging and solvent extraction apparatus (SPE) developed by Umano and Shibamoto (*11*). The headspace of the brewed coffee mixture was purged into 250 mL of deionized water with a purified nitrogen stream while the mixture (10 mL/min) was stirred at 60 °C; the water solution was extracted with 50 mL of dichloromethane simultaneously and continuously for 3 h. The extract was dried over anhydrous sodium sulfate for 12 h. After removal of the sodium sulfate, the extract was concentrated using fractional distillation, and subsequently further concentrated under a nitrogen stream to 1 mL.

The concentrated sample was placed in a glass column (15 cm X 1 cm i.d.) packed with silica gel (Kieselgel 60, E. Merck, Darmstadt, Germany). The sample was sequentially developed into five fractions with a 30 mL solvent mixture of different ratios of pentane and ethyl acetate (100/0, 100/0, 95/5, 5/95, 0/100). Each fraction was concentrated to a final volume of 0.5 mL by fractional distillation and stored at −5 °C for subsequent experiments. The experiment was replicated four times.

Sample Preparation with Liquid-Liquid Continuous Extraction (LLE). Ground roasted regular or decaffeinated coffee (75 g) was brewed with 600 mL tap water. The freshly brewed coffee was extracted with 100 mL dichloromethane using a liquid-liquid continuous extractor for 6 h. The extract was dried over anhydrous sodium sulfate for 12 h. After removal of the sodium sulfate, the extract was concentrated using fractional distillation, and subsequently further concentrated under a nitrogen stream to 1 mL.

The concentrated sample was placed in a glass column (15 cm X 1 cm i.d.) packed with 160–200 mesh silica gel (J. T. Baker Inc., NJ) and the sample was sequentially developed into seven fractions with a 100 mL solvent mixture of different ratios of pentane and ethyl acetate (100/0, 95/5, 80/20, 50/50, 20/80, 0/100) and finally with 200 mL acetone. Each fraction was concentrated to a final volume of 1 mL by fractional distillation and stored at –5 °C for subsequent experiments. The experiment was replicated four times. These experiments were simultaneously performed with controls that contained aldehydes and a GC internal standard only.

Antioxidation Test. Antioxidative activity of the samples was tested using their inhibitory effect toward conversion of aldehyde to acid (*12*). A testing sample (100 mL) was added to a 1 mL dichlomethane solution of pentanal (1 mg/mL) containing 0.1 mg/mL of 2,5-dimethyl hexane as a gas chromatographic internal standard or a 1 mL dichloromethane solution of hexanal (3 mg/mL) containing 0.1 mg/mL of undecane as a gas chromatographic internal standard. The oxidation of the sample solution was initiated by bubbling air at 70 °C for 1 min. The increase in acid or decrease in aldehyde was monitored at 2-day time intervals.

The authentic chemicals of caffeine, 5-HMF, maltol, BHT, and a-tocopherol were also examined for their antioxidative activity using this same testing method.

Quantitative Analysis of Aldehydes and Acids. The quantitative analysis was conducted according to the internal standard method. A Hewlett-Packard (HP) model 5790 gas chromatograph (GC) equipped with a 60 m X 0.25 mm i.d. DB-5 bonded-phase fused-silica capillary column (J & W Scientific, Folsom, CA) and a flame ionizatio detector (FID) was used to monitor the relative amounts of pentanal present in the samples prepared by SPE. The injector temperature was 250 °C and the detector temperature was 300 °C. The oven temperature was programmed from 50 °C to 200 °C at 6 °C/min.

An HP model 5890 GC equipped with a 30 m X 0.25 mm i.d. DB-1 bonded-phase fused-silica capillary column (J & W Scientific, Folsom, CA) and an FID was used for the samples obtained by liquid-liquid continuous extraction. The injector and detector temperatures were 250 °C and 275 °C, respectively. The oven temperature was held at 40 °C for 5 min and then was programmed to 180 °C at 3 °C/min.

An HP 5890 series II gas chromatograph interfaced to an HP 5971 A mass selective detector (GC/MS) was used for mass spectral identification of the GC components at MS ionization voltage of 70 eV. Column and oven conditions were as stated above.

Results and Discussion

The aldehyde/carboxylic acid test is a fast and simple method to assess the antioxidative properties of chemicals or groups of chemicals. Fatty aldehydes are converted readily to a corresponding fatty acid in the oxygen-rich dichloromethane solution through a radical-type reaction (*13*).

Figure 1 shows the relative peak area of pentanal for column chromatographic fractions prepared from the extract of SPE and for the control throughout a storage period of 10 days. Fraction III which was eluted with a pentane/ethyl acetate (95/5) solution inhibited the aldehyde/carboxylic acid conversion for more than 10 days. All fractions except fraction III exhibited weak antioxidative activity. 1-Methylpyrrole and 1-furfurylpyrrole were identified in fraction III. 1-Methylpyrrole, which was also found in the headspace of the heated corn oil/glycine model system, inhibited conversion of pentanal to pentanoic acid for more than 80 days at a level of 500 mg/mL (*12*). The five-membered heterocyclic aromatic ring is reportedly able to scavenge reactive radicals (such as a hydroxy radical) (*14, 15*). As hypothesized in Figure 2, 1-methylpyrrole formed a hydroxy adduct with a hydroxy radical. The water elimination step may not occur under neutral pH (*14*).

Many pyrroles have been reported in numerous heat-treated foods such as cooked meats, roasted beans and nuts, and brewed coffee (*16, 17*).

In order to validate the testing method, activities of known antioxidants, a-tocopherol and BHT, were measured in a hexanal/hexanoic acid system. The activity was monitored by measuring amount of hexanoic acid formed. Figure 3 shows the results of the antioxidative activity test on a-tocopherol and BHT (along with maltol, 5-HMF, and caffeine). Both antioxidants exhibited dose-related activity. They inhibited hexanoic acid formation by 100% at the level of 250 mg/mL; at the level of 8 mg/mL, both antioxidants inhibited the acid formation by 50%. BHT inhibited hexanoic acid formation by 100% at the level of 50 mg/mL, whereas a-tocopherol inhibited the hexanoic acid formation by 90% at the same level. These results are consistent with previous reports (*7, 13*). The results indicate that this method is useful for examining the antioxidative activity of chemicals.

Figure 4 shows the results of the antioxidative test on fractions obtained from a dichloromethane extract of regular brewed coffee using a liquid-liquid extraction method. All fractions except Fraction I (100% pentane eluate) inhibited the acid formation by almost 100% over 14 days. In particular, Fraction VII (100% acetone eluate) exhibited strong antioxidative activity. Hexanoic acid contained originally in the testing solution of hexanal disappeared by the action of Fraction VII after 4 days, suggesting that this fraction contained some reducing agents. Among the 26 compounds identified in Fraction VII, maltol and 5-HMF were tested for antioxidative activity. The results are shown in Figure 3. Maltol inhibited hexanoic acid formation by 100% at levels higher than 250 mg/mL. Dose-related inhibitory activity was observed in the case of 5-HMF which inhibited hexanoic acid formation by 95% and 50% at the levels of 500 mg/mL and 50 mg/mL, respectively. Hypothesized mechanisms of a hydroxy radical abstraction with maltol and 5-HMF based on the previous reports (*14, 15, 18, 19*) are shown in Figure 5.

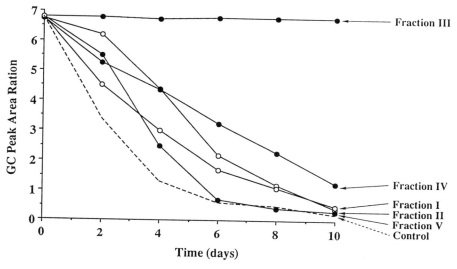

Figure 1. Relative amounts of remaining pentanal in samples containing column chromatographic fractions from headspace of brewed coffee throughout a storage period of 10 days.

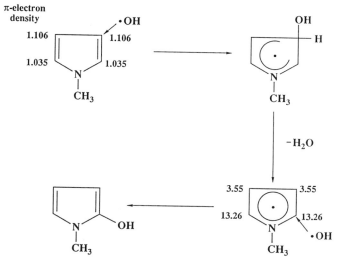

Figure 2. Hypothesized mechanisms of a hydroxy radical abstraction by N-methylpyrrole.

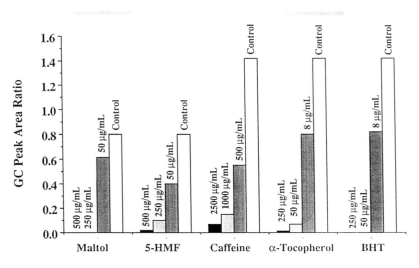

Figure 3. Relative amounts of hexanoic acid formed from hexanal in samples containing antioxicants and control after 10 days of storage.

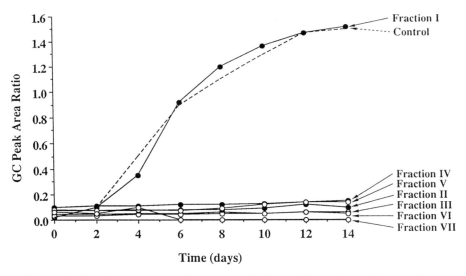

Figure 4. Relative amounts of hexanoic acid formed from hexanal in samples containing column chromatographic fractions from a dichloromethane extract of brewed regular coffee throughout a storage period of 14 days.

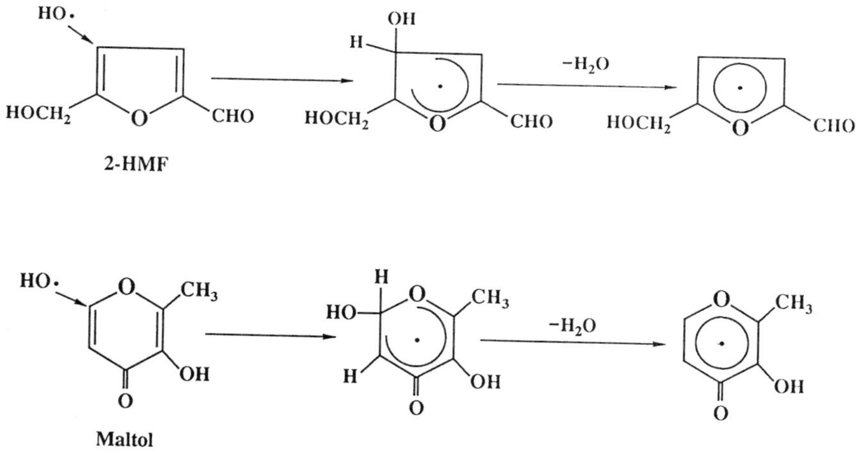

Figure 5. Hypothesized mechanisms of a hydroxy radical abstraction by maltol and 5-HMF.

Caffeine comprised approximately 70% of Fraction VI (100% ethyl acetate eluate). The results of the antioxidative test on authentic caffeine is also shown in Figure 4. Caffeine inhibited the acid formation by 90% at the level of 2.5 mg/mL. Caffeine's antioxidative activity was much less than that of either maltol or 5-HMF. In the case of decaffeinated coffee (Figure 6), all fractions except Fraction I inhibited hexanoic acid formation by almost 100%. However, Fraction I exhibited appreciable inhibitory activity. The results suggest that the antioxidative activities of coffee extracts are due not only to the presence of caffeine, but also to the presence of other antioxidants such as volatile heterocyclic compounds.

Antioxidative compounds such as chlorogenic acid, caffeic acid, quinic acid, and furulic acid have been reported in coffee beans (20). Those phenolic acids were found in greater amounts in green coffee beans and were present in significantly decreased amounts—up to 50-90%—in roasted coffee (21). The overall antioxidative activity of roasted coffee is due not only to the presence of phenolic acids but also—more importantly—to the presence of Maillard reaction products generated during heating processes such as roasting and brewing (22).

Coffee is one of the most popular beverages in the world. Its unique flavor has been studied intensively and constantly since the beginning of this century. The number of volatile chemicals identified in coffee has reached almost 1000 (10). The coffee components that received the most attention from flavor chemists are heterocyclic compounds because of their characteristic roasted or toasted flavor. The number of heterocyclic compounds—including thiophenes, thiazoles, oxazoles, pyrroles, pyrazines, and furans—found in coffee is almost 350 (23). Recently, antioxidative activities of the above mentioned heterocyclic compounds have been reported. For example, alkylthiophenes, 2-thiophenethiol, 2-methyl-3-furanthiol, furfuryl mercaptan, 2-thiothiazoline, and imidazole—which were all found in coffee—inhibited hexanal oxidation for up to 30 days (7, 9). These compounds also

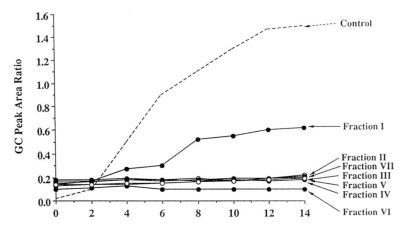

Figure 6. Relative amounts of hexanoic acid formed from hexanal in samples containing column chromatographic fractions from a dichloromethane extract of brewed decaffeinated coffee throughout a storage period of 14 days.

exhibited antioxidative activities measured in lipid peroxidation systems and by the thyrosyl radical scavenging assay (*24*).

The presence of various heterocyclic compounds may explain the improvement of food stability. Moreover, ingestion of these heterocyclic compounds may help to prevent *in vivo* oxidative damage such as lipid peroxidation which is associated with many diseases, including cancer, arteriosclerosis, aging, and diabetes.

Literature Cited

1. Griffith, T.; Johnson, J. A. Cereal Chem. **1957**, *34*, 159–169.
2. Yamaguchi, N.; Yokoo, Y.; Koyama, Y. Nippon Shokuhin Kogyo Gakkaishi **1964**, *11*, 184–189.
3. Boon, P. M. New Zealand J. Dairy Sci. Technol. **1976**, *11*, 278–280.
4. Hauri, J. *Der Einfluss verschiedener Herstellungsverfahren auf die Haltbarkeit walzengetrockneter Getreideflocken* (dissertation). ADAG Administration & Druck AG, Zurich, 1982.
5. Zipser, M. W.; Watts, B. M. Food Technol. **1961,** *15*, 445–447.
6. Yamaguchi, N. In *Proceedings of the 3rd International Symposium on the Maillard Reaction.* Fujimaki, M.; Namiki, M.; Kato, H. Eds.; Elsevier: Amsterdam, **1986**; pp 291-299.
7. Shaker, E. S.; Ghazy, M. A.; Shibamoto, T. *J. Agric. Food Chem.* **1995**, *43*, 1017–1022.
8. Shibamoto, T. In *Instrumental Analysis of Foods;* Charalambous, G., Inglett, G., Eds.; Academic Press: New York, NY, Vol. I. **1983**; pp 229–278.
9. Eiserich, J. P.; Shibamoto, T. *J. Agric. Food Chem.* **1994**, *42*, 1060–1063.
10. Shibamoto, T. In *Proceedings of the 14th International Scientific Colloquium on Coffee, San Francisco, 1991*; Association Scientifique Internationale du Cafe: Paris, **1992**, pp 107–116.
11. Umano, K.; Shibamoto, T. *J. Agric. Food Chem.* **1987**, *35*, 14–18.
12. Macku, C.; Shibamoto, T. *J. Agric. Food Chem.* **1991**, *39*, 1990–1993.
13. Nonhebel, D. C.; Tedder, J. M.; Walton, J. C. *Radicals;* Cambridge University Press: London, **1979**; p 157.
14. Samuni, A.; Neta, P. *J. Phys. Chem.* **1973**, *77*, 1629–1635.
15. Mahanti, M. K. Ind. *J. Chem.* **1977**, *15B*, 168–174.
16. Vitzthum, O. G.; Werkhoff, P. *J. Food Sci.* **1974**, *39*, 1210–1215.
17. Tressl, R.; Silwar, R. *J. Agric. Food Chem.* **1981**, *29*, 1078–1082.
18. Shiga, T.; Isomoto, A. *J. Phys. Chem.* **1969**, *73*, 1139–1143.
19. Schuler, R. H.; Laroff, G. P.; Fessenden, R. W. *J. Phys. Chem.* **1973**, *77*, 456–466.
20. Stich, H. F. *Mutation Res.* **1991**, *259*, 307–324.
21. Clifford, M. N. *Food Chem.* **1979**, *4*, 63–71.
22. Kato, H.; Lee, I. E.; van Chunyen, N.; Kim, S. B.; Hayase, F. *Agric. Biol. Chem.* **1987**, *51*, 1333–1338.
23. Flament, I.; Chevalier, C. *Chem. Ind.* **1988**, 592-596.
24. Eiserich, J. P.; Wong, J. W.; Shibamoto, T. *J. Agric. Food Chem.* **1995**, *43*, 647–650.

MEDICINAL PLANTS

Chapter 12

Garlic and Related *Allium* Derived Compounds: Their Health Benefits in Cardiovascular Disease and Cancer

Manfred Steiner and George Sigounas

Division of Hematology/Oncology, East Carolina University School of Medicine, Greenville, NC 27858–4354

Garlic and garlic extracts have been shown to reduce cardiovascular risk factors and carcinogenesis induced by certain chemicals and viruses. Total and LDL cholesterol are lowered by up to 8%, blood pressure by about 5%, but not HDL cholesterol or serum triglycerides. Platelet adhesion to fibrinogen was lowered by aged garlic extract by 30%, platelet aggregation induced by epinephrine by up to 80% and with collagen as stimulant a 25% reduction was noted. Induction of cancer by a number of chemical carcinogens could be inhibited by several thioallyl compounds derived from garlic and also H-*ras* oncogene transformed tumor growth was inhibited by an organosulfide from garlic. The biological effects of garlic can be attributed primarily to its organosulfur compounds.

Allium vegetables, most notably *allium sativum*, have been believed for thousands of years to carry health benefits for those who regularly consume it. While much of this appears to be folklore, some of the benefits attributed to garlic appear to be founded on experimental evidence. In the last two decades, increasing attention has been paid to the biologic functions and medicinal value of naturally occurring compounds. There exists a general belief that such substances are better tolerated and less toxic than synthetic compound administered as pharmaceutical.

The substances held responsible for the biological and medicinal functions of garlic belong to a large group of thioallyls (1,2). Major representatives of this group of compounds are shown in Figure 1. The majority of these organosulfurs are oil soluble but hydrophilic compounds, e.g. cysteine derivatives and γ-glutamylcysteines are also represented in garlic. In addition, a wide variety of other substances such as amino acids especially arginine, allixin, polysaccharide, saponins and a sizable protein faction are present in garlic extracts or garlic some of which may have biological functions. A host of thioallyl substances arises from the conversion of alliin by alliinase to allicin (2). This enzyme becomes activated when garlic cloves are crushed or cut. Allicin and allicin potential, that is alliin and alliinase activity, have been considered as indices for evaluating and comparing the medicinal value of garlic preparations commercially sold.

The stability of this compound varies widely depending on the milieu it is dissolved in (*3*). At acid pH, allicin is quite stable, but it does not generate any of its normal transformation products at such pH values. In the blood, especially in red cells, allicin disappears extremely rapidly and exposure of hemoglobin to this thioallyl compound produces significant amounts of methemoglobin a hemoglobin that is incapable of participating in oxygen transport (*3*). Thus it is not yet clear to what extent such unstable thioallyl compounds contribute to the presence of reactive organosulfurs in tissues. The stable thioallyl compounds S-allylcysteine (SAC) and S-allylmercaptocysteine (SAMC) are primarily present in aged garlic extract. With a number of companies that are preparing and selling various garlic extracts or garlic preparations in the market place, it has become quite difficult to distinguish between the different claims made for the various products. For that reason we shall restrict our remarks on the biological efficacy of garlic and or its active principals to those for which the scientific basis appears to be sound and if possible confirmed by other independent studies.

I *Cardiovascular Effects of Garlic or Garlic Extracts*

Among the multiple risk factors for the development of cardiovascular disease (*4*) several have been shown or claimed to be influenced by the administration of garlic and or its extracts. The one most commonly cited as being lowered by garlic is hyperlipidemia, specifically hypercholesterolemia. A large number of clinical intervention trials both in subjects having normal serum cholesterol levels as well as those showing hypercholesterolemia using a variety of commercial garlic preparations have been published. Two meta-analyses have attempted to sift through these studies and eliminate those that were flawed from consideration (*5,6*). Although widely varying dosages of garlic extracts have been applied in these trials, a distinct but moderate cholesterol lowering effect has been detected. LDL but not HDL cholesterol were shown to be reduced and levels of serum triglycerides variably reduced or left unaltered by administration of garlic extracts. The observed reductions in total cholesterol ranged from about 5 to 12% with an average of 8%. In our own studies (*7*), using aged garlic extract at a supplementation level of 7.2 g/day, a modest but statistically significant reduction in total cholesterol as well as LDL cholesterol was recorded in a long term administration study of mildly hypercholesterolemic men (Fig. 2).

The reason why garlic should reduce cholesterol levels is not completely clear. There are only very few studies that focused on the mechanism of action. One theory postulated inhibition of a crucial enzyme in the biosynthesis of cholesterol, i.e. 3-hydroxy-3-methylglutaryl-CoA reductase (*8,9*). This enzyme is also the target of most of the currently effective cholesterol lowering pharmaceuticals. In tissue culture experiments reductions of the enzyme activity have been reported (*8,10*). Whether this is the only mechanism responsible for lowering cholesterol is unknown at this time.

Hypertension is another prominent risk factor for the development of cardiovascular disease that appears to be influenced in some measure by the administration of garlic or garlic extract. There are as yet no studies that have addressed the possible mechanism of the decrease in blood pressure due to garlic preparations. The aged garlic extract used in our studies (Kyolic) (*7*) led to a moderate decrease in systolic blood pressure (Fig. 3). This extract contains fructosyl-arginine. It is possible

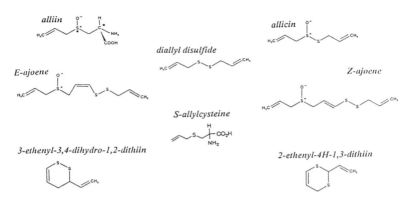

Figure 1: Major organosulfur compounds present in garlic.

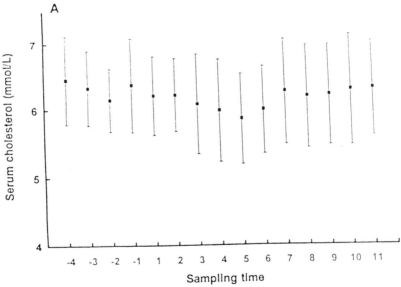

Sampling time

Figure 2: Serum total cholesterol concentrations of subjects in a 11 month double blind cross-over study of 41 moderately hypercholesterolemic men (4 week baseline evaluations -1 to -4). A) Subjects starting their intervention trial with 7.2 g aged garlic extract/day (sampling time 1-6) and B) subjects beginning their intervention trial with placebos followed by aged garlic extract (sampling time 7-11). Within and across study arm comparisons showed all highly significant differences. Reproduced with permission from ref. 7. Copyright 1997 American Society for Clinical Nutrition.

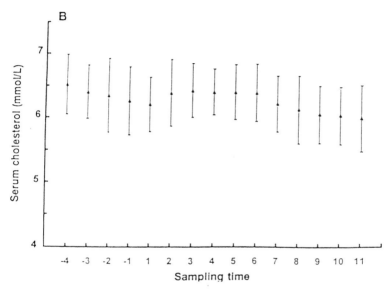

Figure 2. *Continued.*

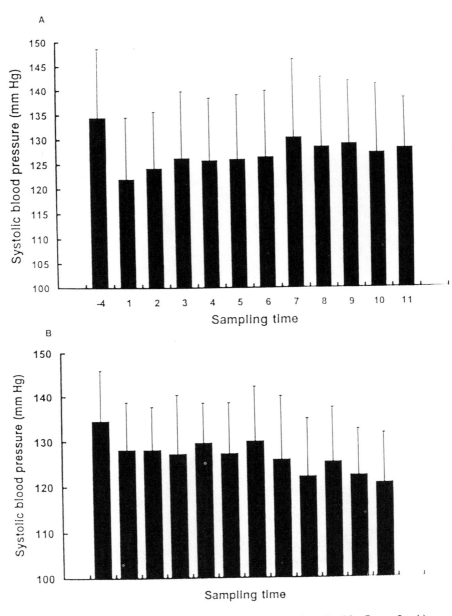

Figure 3: Systolic blood pressure readings in subjects described in figure 2. A) Subjects beginning their intervention trial on aged garlic extract (sampling time 1-6). B) Subjects on placebos during study arm 1 and on aged garlic extract during the second arm of the study (sampling time 7-11). Reproduced with permission from ref. 7. Copyright 1997 American Society for Clinical Nutrition.

that this substance is a substrate for NO synthase. Stimulation of the enzyme could have lowering effect on blood pressure. Other investigators found decreases in blood pressure (11,12), most of them in the same range as the reduction obtained in our study (7). Diastolic pressure also showed a reduction, but it was not as pronounced as that of systolic blood pressure.

Hypercoagulability is a potential risk factor for cardiovascular disease. As platelets are prominently involved in the pathogenetic mechanism of thrombosis in the arterial circulation, e.g. coronary vasculature and cerebral blood vessels, agents that inhibit their function have potential usefulness in reducing cardiovascular disease. There are two fundamentally different platelet functions of importance for physiological and pathological events related to hemostasis and thrombosis, aggregation and adhesion. The former denotes a sticking of platelets to each other when activated by turbulent flow conditions or stimulation by agonists such as ADP, epinephrine, collagen etc. and adhesion, an attachment of platelets to a surface area - usually collagen - which becomes exposed when endothelial cells that line blood vessels are lost for a variety of reasons. A large number of studies have addressed the effect of various garlic preparations, extracts of garlic and individual organosulfur compounds found in garlic on platelet aggregation (13-17). Most of these studies were carried out in vitro and show inhibitory effect of various allylsulfides as well as ajoene and other thioallyl compounds on platelet activation by a variety of agonists. Adhesion of platelets has not been studied extensively, probably because of the more difficult analysis of this platelet function. Although the results were variable, inhibition of platelet adhesion and thrombus formation has been found (18-19).

We have performed platelet function studies in a group of mildly hypercholesterolemic men who were enrolled in a double blind cross-over investigation of aged garlic extract (7.2 g/day) vs placebo (Steiner, M. East Carolina University, unpublished data). The aged garlic extract produced a strong inhibition of epinephrine-induced aggregation of platelets and a moderate decrease in collagen-induced aggregation (Fig. 4). Adhesion of platelets to fibrinogen coated surfaces was also significantly reduced (about 30%) in response to aged garlic extract administration.

The mechanism for these effects of the garlic extract are not completely clear, but our studies have revealed a decrease in acid precipitable free sulfhydryl groups in platelets exposed to SAMC or SAC. Although the concentrations used in the in vitro experiments were considerable higher than could be achieved by dietary supplementation, we believe that in vivo allylsulfides accumulate as final reaction products of the thioallyl compounds in garlic extract as they are lipid soluble and thus presumably have a longer biological half life. It is possible that thioallyl compounds interfere with the membrane localization of rap- or rac-related proteins which participate in the response of platelets to agonists (20). Blocking of the C-terminal cysteine residues required for prenylation of such proteins could be a potential mechanism of inhibition. There are also a number of accessible SH groups in the major glycoprotein present in platelet membranes, i.e. α_{gpIII}/β_3 integrin (21). The heterodimer gp I/IX, the receptor for von Willebrand factor that allows platelets to interact with collagen in the subendothelial tissues also has accessible cysteinyl residues (21) that may interact with thioallyl compounds. Thus there are multiple points for potential interference that may play a role in the aged garlic extract induced inhibition of platelet functions.

From this description of the effects of garlic and garlic extracts on several risk

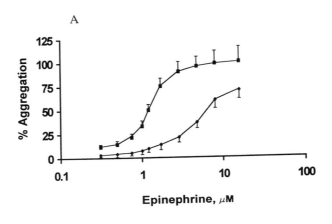

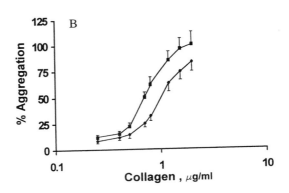

Figure 4: Reduction in platelet aggregation by epinephrine (A) and collagen (B) in a subgroup of the study patients described in figure 2. Mean ± SD of 10 measurements are shown. Statistically significant differences between placebo and aged garlic extract (7.2 g/day) were noted for epinephrine and collagen.

factors for cardiovascular disease development it is apparent that this dietary supplement has unique properties in being able to interfere with several of these risk factors. Although individually the effect of garlic and its extracts may not be as prominent as that of pharmaceuticals specifically directed against hypercholesterolemia, hypertension and platelet functions, it is the combination of these inhibitory effects in a single preparation that is of special significance.

II *Anticarcinogenic Effects of Garlic*

Prompted by a rich anecdotal history of beneficial effects of garlic on the incidence of cancer and its development, investigators have studied the action of the major sulfur compounds present in garlic on chemical and to a lesser extent on viral induced carcinogenesis. Diallylsulfide was found to strongly inhibit certain cytochrome P450 enzymes (P450IIE1) which are essential for the oxidative activation of polycyclic aromatic hydrocarbons, e.g. benzo[*a*]pyrene (*22*). The thioallyl compound was found to have high affinity for the enzyme, thus strongly inhibiting it. Immunologically reactive protein was also decreased by diallylsulfide. The effect was specific for P450IIE1 as the opposite effect was observed with P450IIB1. Increased levels of glutathione-S-transferase (GST) were noted by several investigators (*23,24*) in a variety of tissues in response to several thioallyl compounds administered to experimental animals. This enzyme is of preeminent importance in the detoxification of many xenobiotics in the body (*25*). Increased amounts of the various isoforms in specific tissues should increase the conjugation and excretion of carcinogens and their metabolic conversion products. This effect of thioallyl compounds on GST activity was tissue specific and depended on the type of thioallyl compound used (*23,26*). Cellular levels of glutathione and acid soluble free sulfhydryl groups in general showed a variable response to garlic derived organosulfurs (*27*). S-allylcysteine, a compound represented especially in aged garlic extract initially raised levels of free SH groups in a variety of malignant and non-malignant cells but subsequently on continued administration for longer periods of time produced a dose-dependent decrease compared to appropriate controls (Fig. 5). Structural consideration of SAC can readily explain this effect as being due to the release of cysteine from SAC. The lowering of free SH groups is probably due to progressive alkylation of SH groups by allylsulfides. This latter effect may be cumulative as allylsulfides are less water soluble than cysteine.

A variety of other biochemical abnormalities have been detected in tumors or isolated cell line exposed to organosulfurs derived from garlic. Decreased DNA methylation (*28*), reduction of ornithine decarboxylase activity (*29*) and decrease in aflatoxin adduct formation (*30*) to name but a few.

Viral carcinogenesis was also found to be suppressed by thioallyl compounds specifically dialylldisulfide (*10*). H-ras oncogene transformation and tumor growth was inhibited apparently by blocking the membrane association of the ras oncoprotein. Inhibition of 3-hydroxy-3-methylgultaryl CoA-reductase activity was thought to be responsible for this effect. Although farnesyl transferase activity was unaffected by the dialylldisulfide treatment, one cannot rule out a possible effect of thioallyl compounds on the C-terminal cysteine residues of ras protein that are required for the prenylation of such proteins.

120

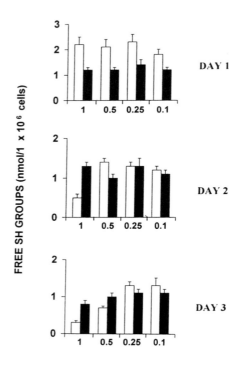

CONCENTRATIONS OF ADDITIVES (mM)

Figure 5: Quantitative analysis of acid-soluble free SH groups in erythroleukemia cells (HEL) grown in media supplemented for up to 3 days with different concentrations of SAMC (open bars) or equal volumes of solvent (solid bars). Means ± SD of 3 experiments. Reproduced with permission from ref. 27. Copyright 1997 Lawrence Erlbaum Associates, Inc.

S-allylmercaptocysteine induced apoptosis in a number of tumor cell lines including breast and prostate cancer (*27*). The mechanism of this effect has not yet been elucidated but interleukin-1β converting enzyme could be play a role (*31*). The latter is a cysteine protease the activity of which could be stimulated by increased levels of free sulfhydryls in the cell.

Cell proliferation is inhibited by SAMC with the cell cycle being arrested in G2/M phase (Fig. 6) (*27*). Thymidine incorporation was strongly inhibited by the same compound in a variety of proliferating cells. Not only malignant cell lines but also

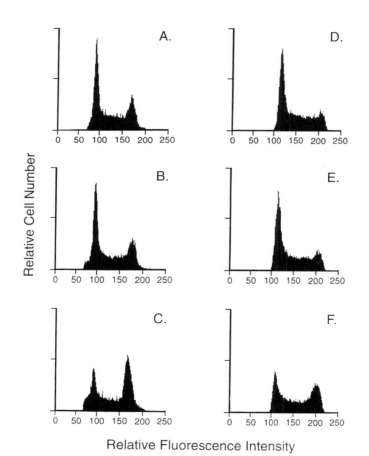

Figure 6: Effect of SAMC on erythroleukemia cell cycle progression. HEL (A-C) and OCIM-1 cells (D-F) were grown up to 3 days in media supplemented with 0.1 mM SAMC (C and F), an equivalent volume of solvent (B and E), or without additive (A and D). After cells were stained with propidium iodide, DNA histograms were obtained by flow cytometry. Reproduced with permission from ref. 27. Copyright 1997 Lawrence Erlbaum Associates, Inc.

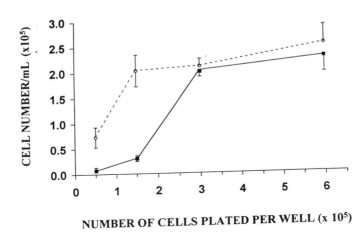

Figure 7: Proliferation of confluent and nonconfluent human umbilical vein endothelial cells in response to SAMC. Different concentrations of endothelial cells supplemented with 0.25 mM SAMC (solid line) or equal volumes of solvent (dashed line) were plated and grown for 24 h. Means ± SD of 3 experiments. Reproduced with permission from ref. 32. Copyright 1997 Lawrence Erlbaum Associates, Inc.

normal proliferating cells were inhibited in their growth by such compounds (*32*). However, when normal cells reached confluence the growth inhibitory effect as well as the apoptosis inducing action of thioallyl compounds by-and-large ceased (Fig.7).

These results demonstrate that garlic derived organosulfur compound not only affect carcinogenesis but also have the ability to slow the growth of established cancer cells. Epidemiological investigations seem to corroborate at least some of the experimental studies. Although it is difficult to accurately assess the effect of a food supplement that is primarily consumed as seasoning rather than as a staple food, there are a number of large studies that have shown a beneficial effect with significant reduction in the incidence of stomach (*33,34*), colon (*35*), and laryngeal cancer (*36*) in persons who have a high intake of garlic compared to those who do not. A large epidemiological study from the Netherlands (*37*) showed no correlation between garlic intake and incidence of lung cancer but increased onion consumption, *allium* cepa, correlated with a reduction of this type of cancer.

There is mounting evidence that garlic and its thioallyl compounds may have beneficial effect on the development of cancers, especially as a cancer preventive dietary supplement. One has to be careful however, to generalize from in vitro and animal experiments to humans. In the former group of experiments by-and-large single organosulfur compounds were tested at concentrations that exceeded those present in garlic or garlic extract when administered to humans. The absence of side effects when taking garlic or garlic extract does not necessarily extend to the intake of isolated components of organosulfurs. The latter are potent compounds that affect primarily sulfur metabolism, and thus their effect is widespread as demonstrated by the variety of biological functions affected. Nevertheless, we believe that there is considerable therapeutic and preventive potential in thioallyl compounds that could be of considerable health benefit to those who regularly consume such compounds in form of allium vegetables.

Acknowledgments

This study was in part supported by National Heart, Lung, and Blood Institute Grant HL-39109 from the National Institutes of Health.

Literature Cited

1) Lin, R.I. In *Functional Foods*; Editor, I. Goldberg; Chapman & Hall Publ.: New York, NY, 1994, pp 393-449.
2) Block, E. *Sci. Am.* **1985**, *252*, 114-119.
3) Freeman, F., Kodera, Y. *J. Agric. Food Chem.* **1995**, *43,* 2332-2338.
4) Farmer, J.A., Gotto, A.M. In Heart Disease. A Textbook of Cardiovascular Medicine; Braunwald, E., ed.; W.B. Saunders Co., Philadelphia, PA, 4th edition, 1992, pp 1125-1160.
5) Warshafsky, S., Kamer, R.S., Sivak, S.L. *Ann. Int. Med.* **1993**, *119,* 599-605.
6) Silagy, C., Neil, A. *J. R. Coll. Physicians Lond.* **1994**, *28*, 39-45.
7) Steiner, M., Khan, A.H., Holbert, D., Lin, R.I. *Am. J Clin. Nutr.* **1996**, *64*, 866-870.
8) Gebhardt, R. *Lipids*, **1993**, *28*, 613-619.
9) OmKumar, R.V., Banerji, A., RamaKrishna Kurup, C.K., Ramasarma, R. *Biochim. Biophys. Acta,* **1991**, *1078*, 219-225.
10) Singh, S.V., Mohan, R.R., Agarwal, R.,Benson, P.J., Hu, X., Rudy, M.A., Xia, H., Katoj, A., Srivastava, S.K., Mukhtar, H., Gupta, V., Zaren, H.A. Biochem. Biopys. Res. Commun. **1996**, *225,* 660-665.
11) Barrie, S.A., Wright, J.V., Pizzorno, J.E. J. Orthomol. Med. 1987, *2*, 15-21.
12) Kandziora, J. *Arzneimittelforschung* **1988**, *1*, 1-8.
13) Bordia, A. *Atherosclerosis* **1978**, *30*, 355-360.
14) Vanderhoek, J.T., Makheja, A.N., Bailey, J.M. *Biochem. Pharmacol.* **1980**, *29*, 3169-3173.
15) Ariga, T., Oshiba, S., Tamada, T. *Lancet* **1981**, *1*, 150-151.
16) Mohammad, S.F., Woodward, S.C. *Thromb. Res.* **1986**, *44*, 793-806.
17) Ali, M., Mohammed, S.Y. *Prost. Leuk. Med.* **1986**, *25*, 139-146.
18) De Boer, L.W.V., Folts, J.D. *Am. Heart J.* **1989**, *117*, 973-975.
19) Apitz-Castro, R., Badimon, J.J., Badimon, L. Thromb. Res. **1994**, *75*, 243-249.
20) Farrell, F., Torti, M., Lapetina, E.G. *J. Lab. Clin. Med.* **1992**, *120*, 533-537.
21) Principles of Cell Adhesion; Richardson, P.D., Steiner, M., eds.; CRC Press: Boca Raton, FL, 1994.
22) Brady, J.F., Li, D., Ishizaki, H., Yang, C.S. *Cancer Res.* **1988**, *48*, 5937-5940.
23) Sumiyoshi, H., Wargovich, M.J. *Cancer Res.* **1990**, *50*, 5084-5087.

24) Takada, N., Matsuda, T., Otoshi, T., Yano, Y., Otani, S., Hasegawa, T., Nakae, D., Konishi, Y., Fukushima, S. *Cancer Res.* **1994**, *54*, 2895-2899.

25) Wilce, M.C.L., Parker, M.W. *Biochim. Biophys. Acta* **1994**, *1205*, 1-18.

26) Sparnins, V.L., Barany, G., Wattenberg, L.W. *Carcinogenesis* **1988,** *9*, 131-134.

27) Sigounas, G., Hooker, J.L., Li, W., Anagnostou, A., Steiner, M. *Nutrition & Cancer* **1997**, *28*, 153-159.

28) Ludeke, B.I., Domine, F., Ohgaki, H., Kleihues, P. *Carcinogenesis* **1992**, *13*, 2467-2470.

29) Hu, P.J., Wargovich, M.J. *Cancer Lett.* **1989**, *47*, 153-158.

30) Tadi, P.P., Lau, B.H., Teel, R.W., Herrmann, C.E. *Anticancer Res.* **1991**, *11*, 2037-2041.

31) Patel, T., Gores, G.J., Kaufmann, S.H. *FASEB J.* **1996**, *10*, 587-597.

32) Sigounas, G., Hooker, J., Anagnostou, A., Steiner, M. *Nutrition & Cancer* **1997**, *27*, 186-191.

33) You, W.C., Blot, W.J., Chang, Y.S., Ershow, A., Yang, Z.T., An, Q., Henderson, B.E., Fraumeni, J.F. *J. Natl. Cancer Inst.* **1989**, *81*, 162-164.

34) Buiatti, E., Palli, D., Decarli, A., Amadori, D., Avellini, C., Bianchi, S., Biserni, R., Cipriani, F., Cocco, P., Giacosa, A., Marubini, E., Puntoni, R., Vindigni, C., Fraumeni, J.F., Blot, W. *Internatl. J. Cancer* **1989**, *44*, 611-616.

35) Steinmetz, K.A., Kushi, L.H., Bostick, R.M., Folsom, A.R., Potter, J.D. *Am. J. Epidemiol.* **1994**, *139*, 1-15.

36) Zheng, W., Blot, W.J., Shu, X.O., Gao, Y.T., Ji, B.T., Ziegler, R.G., Fraumeni, J.F. *Am. J. Epidemiol.* **1992**, *136*, 178-191.

37) Dorant, E., van den Brandt, P.A., Goldbohm, R.A. *Cancer Res.* **1994**, *54*, 6148-6153.

Chapter 13

Garlic as a Functional Food: A Status Report

Eric Block

Department of Chemistry, State University of New York at Albany,
Albany, NY 12222

While worldwide sales continue to grow for herbal products
derived from garlic, perhaps the most popular of all herbal cure-
alls, researchers continue to debate their health value and
manufacturers challenge each others' claims. The organosulfur and
organoselenium chemistry of garlic is briefly reviewed. The
"natural flavor" of garlic is not present as such in the intact plant
but is formed upon crushing the cloves. The cloves contain alliin
and related S-alk(en)yl cysteine S-oxides which when co-mingled
with alliinase enzymes produce allicin and other thiosulfinates. The
latter compounds can be further transformed into allyl polysulfides,
ajoene, dithiins, etc. Selenium is found in garlic primarily in the
form of derivatives of selenocysteine. Also discussed is the effect
of processing on garlic constituents, the metabolism of garlic derived compounds, the nature of commercial garlic formulations, and
the health effects of garlic and garlic products.

Worldwide sales of garlic nutrient supplements are impressive in amount, in market
share among other supplements, and in rate of annual increase. In 1993 U.S. garlic
supplement sales amounted to $31.3 million, corresponding to 9.8% (second only to
Echinacea) of total herbal products sales at health food stores. A year later garlic supplement sales had increased 26% to $39.4 million (1,2). Garlic supplement sales in
Germany in 1993 were $100 million, with 8% of the German population estimated to
be regularly consuming such supplements (3). Annual garlic supplement sales in
Canada (1995-6) are estimated to be (in $US) $36 million (30% of the total $120
million herbal remedies market in Canada)(4); analogous sales in Japan are estimated
(by the author) to exceed $100 million. Sales of garlic supplements benefit from
continued favorable hype in the tabloids, whose front pages trumpet "garlic: nature's

miracle healing medicine lowers blood pressure, relieves arthritis pains, guards against major ills" (5), "garlic cures man's cancer ... it really works says doctors" (6), "miracle medicine garlic can make you healthy, happy and help you to live to be 100" (7), and in supermarket "mini-books" titled "Garlic -- the Miracle Herb" (8) and "Garlic, Selenium, Cranberries & Green Tea" (9). A well respected book on herbal medicine, now in its third edition, ranks garlic as the most popular herbal cure-all (10). Claims for the effects of garlic supplements, and other nutritional supplements, are regulated by the FDA. Structure-function (but not disease) claims are allowed if they are supported by a "reasonable" amount of clinical data, indicating that some function in the body can be improved, and if a disclaimer regarding treatment of diseases is included on the label.

In the scientific community, debate continues about the health value of garlic supplements. A recent critical letter in *Lancet* titled "Garlic for flavour, not cardio-protection" (11) drew rebuttals from supporters of the use of garlic products (12-14) along with a reply by the author of the original letter (15). The critical letter notes (11) that a recent "well-designed and executed randomised controlled" study in the UK (16) of the effect of a dried "odor controlled" garlic powder on treatment of hyperlipidaemia failed to show a significant difference in lipid concentrations between treatment and placebo groups after 6 months. A second study in Australia gave similar results (17) while studies in the US did show an effect by garlic products (18). One rebuttal letter suggests that "cardioprotection entails more than lipid reduction," that even with the inclusion of recent studies showing no effect for garlic, the "metareanalysis remained positive," and that the criticism itself may be "a textbook example of bias against complementary/alternative medicine" (12). A second rebuttal letter indicates that studies have shown that "garlic [more precisely, components of garlic extracts] has been shown to inhibit cholesterol biosynthesis in hepatocytes and liver homogenates"; garlic preparations can also affect "an 80% reduction in serum thromboxane concentrations" and can "inhibit platelet aggregation thus reducing the frequency of thrombus formation" (13). In the latter authors' opinion, "the jury is still deliberating on garlic" (13). A third responding letter estimates that "one in three American adults use some form of non-conventional therapy" but then states that "it is essential that the medical and health care communities develop rigour in their understanding and definition of natural products ... Discussion of natural products in medical journals lacks the same rigour ... In phytopharmacology ... one plant species may contain various pharmacologically active constituents and be available as several distinct and separate preparations ... [Therefore] the manufacturing process is the major determinant of a plant's pharma-cological activity" (14). The author of the original letter replies that a positive meta-analysis by itself is not sufficient grounds for supporting the use of garlic for lipid lowering, points to limitations of meta-analyses such as "publication bias with positive studies more likely to be published than negative studies" and states that "more clinical trials are justified" (15). Against the above backdrop, this chapter will summarize the current state of medical and molecular knowledge concerning garlic preparations and their biological activity.

Allium Chemistry

The "natural flavor" of garlic is not present as such in the intact plant but is formed upon crushing or cutting the cloves (*19-22*). When garlic cells rupture, rapid reactions occur when the (separated) flavor precursors and enzymes co-mingle. In intact garlic cloves alliin (compound **1a** in Scheme 1) and related aroma and flavor precursors are located in the abundant storage mesophyll cells while the alliinase enzymes are concentrated in the relatively scarce vascular bundle sheath cells, located around veins or phloem. Alliinase activity is ten times higher in the bulb of a mature plant than in the leaves and stems. This particular arrangement concentrates the enzyme in a region of the garlic bulb where it should be most useful in protecting the bulb against predators or infectious microorganisms through release of natural chemical defense agents. Allicin (**3a**), a thiosulfinate which possesses the characteristic aroma and taste of fresh garlic, is formed by alliinase-induced splitting of alliin (**1a**) into 2-propenesulfenic acid (**2a**). Two molecules of **2a** combine to form one molecule of allicin (ca. 0.37% yield from fresh garlic) plus water.

Alliin is derived from cysteine (**5**; Scheme 2), which is manufactured by garlic from sulfate ions (**4**) found in soil. In addition to alliin, garlic also contains several dozen cysteine-containing peptides such as **6a-d**, as well as three other related flavor precursors **1b-d** which, like alliin (**1a**), are cleaved by alliinases to give thiosulfinate flavorants **3** via sulfenic acids **2** (*19-22*). In Scheme 2, R/R' represent allyl, 1-propenyl, methyl or trace amounts of *n*-propyl. Allicin [RS(O)SR (**3a**), R = allyl (CH_2=CH-CH_2)] makes up about 75% of the garlic thiosulfinates. We have used CO_2 supercritical fluid extraction (*23*) and LC-APCI-MS (liquid chromatography atmospheric pressure chemical ionization-mass spectroscopy) to fully characterize both the major and minor thiosulfinates **3** produced when garlic is crushed (Scheme 3)(*24*). Peptides **6** function as storage compounds of nitrogen and sulfur and can be enzymatically transformed by the plants into precursors **1**. Major peptides in garlic include γ-glutamyl-*S*-2-propenylcysteine (**6a**), γ-glutamyl-*S*-1-propenylcysteine (**6b**), and γ-glutamyl-*S*-methylcysteine (**6c**) (ratio in garlic 40%:58%: 2%)(*25*). The action of γ-glutamyl-transpeptidases (first step) and then oxidases (second step) is required for conversion of these peptides into alliin (**1a**), isoalliin (**1b**), and methiin (**1c**), respectively. Cool storage (wintering) also promotes conversion of peptides to alliin, isoalliin, and methiin. Peptides are only present in bulbs; oxidized peptides are present only in trace amounts in garlic. In garlic cloves **6a** and **6b** are about equally abundant yet alliin (**1a**) is 17 times more abundant than isoalliin (**1b**). This suggests a second predominating biosynthetic route to alliin, other than from γ-glutamyl peptides, e.g. from serine (*21*).

Garlic Organoselenium Compounds. Garlic has a higher sulfur content (0.35%, 3.5 mg/g) than most other common vegetables. The level of the essential micronutrient element selenium in garlic (0.000028%; 0.28 μg/g), although much lower than that of sulfur, is also relatively high compared to other common vegetables. The selenocysteine-derived selenium compounds in garlic are thought to originate from soil selenite (SeO_3^{2-}) or selenate (SeO_4^{2-}). Identification of natural organoselenium

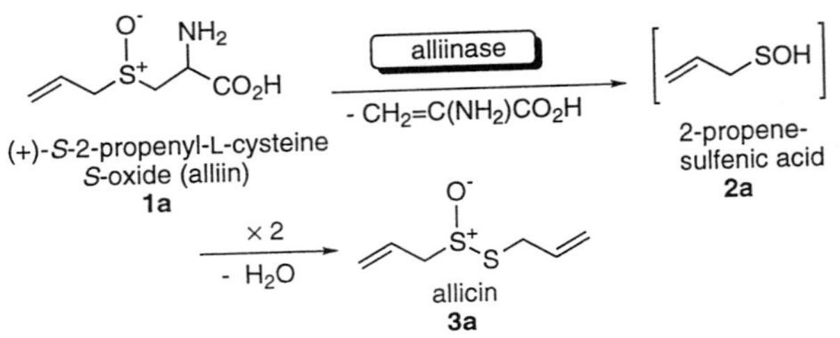

Scheme 1. Mechanism for the formation of allicin upon crushing garlic.

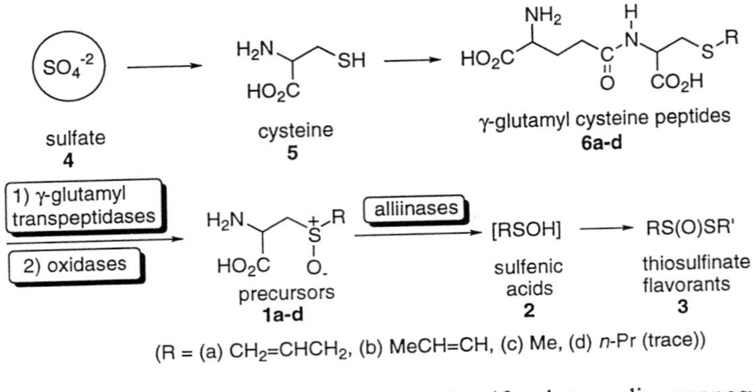

(R = (a) CH$_2$=CHCH$_2$, (b) MeCH=CH, (c) Me, (d) n-Pr (trace))

Scheme 2. Proposal for transformation of soil sulfate into garlic organosulfur
compounds.

compounds is of considerable current interest in view of the discovery that Se-enriched garlic (26-30) as well as Se-enriched yeast (31) possess cancer preventive properties. Detection of natural Se compounds, admixed with far higher levels of chemically similar sulfur compounds, requires the use of element-specific analytical methods such as gas chromatography-atomic emission detection (GC-AED) and high performance liquid chromatography-inductively coupled plasma-mass spectrometry (HPLC-ICP-MS). Using the technique of GC-AED, the Se emission line at 196 nm is monitored to identify organoselenium species while concurrently monitoring S and C by lines at 181 and 193 nm, respectively.

Analysis of the headspace above chopped garlic using GC-AED shows MeS_nMe, MeS_nAll, and $AllS_nAll$ (n = 1-3, All = allyl) in the S channel. The Se channel shows seven peaks: dimethyl selenide (MeSeMe), methanesulfenoselenoic acid methyl ester (MeSeSMe), dimethyl diselenide (MeSeSeMe), bis(methylthio)selenide ($(MeS)_2Se$), allyl methyl selenide (MeSeAll), 2-propenesulfenoselenoic acid methyl ester (MeSeSAll), and (allylthio) (methylthio)selenide (MeSSeSAll) (32-33). Structures were established by GC-MS using synthetic standards. Lyophilized normal garlic (0.02 ppm Se) or moderately Se-enriched garlic (68 ppm Se; grown in Se-enriched soil) was derivatized with ethyl chloroformate to volatize the selenoamino acids, likely precursors of the headspace Se compounds. Analysis by GC-AED showed selenocysteine, identified by comparison with the mass spectral fragmentation and the retention time of an authentic standard. In more heavily Se-enriched garlic (1355 ppm Se), *Se*-methyl selenocysteine is the major selenoamino acid along with minor amounts of selenocysteine and traces of selenomethionine (32,34). HPLC-ICP-MS analysis of high selenium garlic extracts show that *Se*-methyl selenocysteine is the major component along with lesser amounts of selenomethionine, selenocystine, and selenate and selenite salts. A considerable number of unknown selenium-containing components remain to be identified (35). A proposed mechanism for formation of the various selenium compounds thus far identified from garlic is shown in Scheme 4.

Effects of Processing on Garlic Constituents. Conversion of cysteine sulfoxides (**1**) to thiosulfinates (**3**) by alliinases is very fast. It is complete in 10 seconds or less for alliin (**1a**) and 60 seconds for methiin (**1c**). These rates are only moderately increased to 30 seconds and 5 minutes for oven-dried commercial garlic powders (21). Garlic cloves contain equal amounts of alliin and alliinases. Alliinases are robust enzymes, since garlic powder stored for five years shows little loss in allicin formation. The optimium pH for garlic alliinase is 6.5. Garlic appears to have three alliinases, one specific for alliin (**1a**), one for isoalliin (**1b**), a third for methiin (**1c**). The latter enzyme appears to be more heat sensitive than the former two, since heat drying of garlic slices decreases the rate of formation of methyl but not allyl/1-propenyl thiosulfinates. During food preparation involving garlic, additional flavors are formed by thermal breakdown of initial enzymatically-produced flavorants, either in an aqueous or non-aqueous (e.g., cooking oil) medium. Some compounds may be lost by evaporation. Heating in an aqueous medium can lead to hydrolysis products. If the breakdown products themselves are unstable, further compounds can be

130

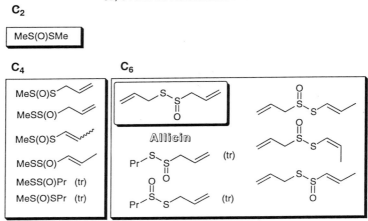

Scheme 3. Major and minor thiosulfinates from crushed garlic.

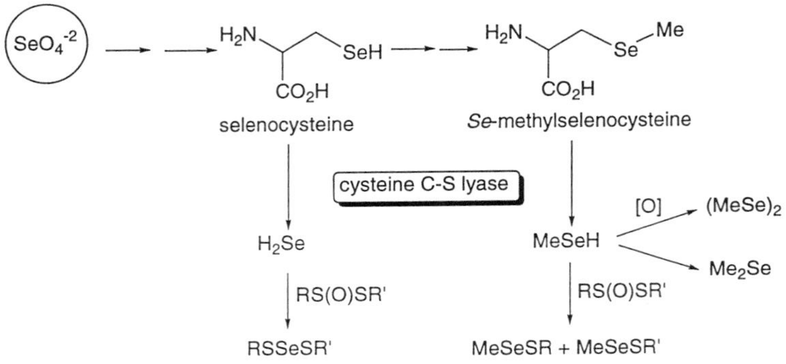

Scheme 4. Proposal for transformation of soil selenate into garlic organoselenium compounds.

formed which may also make important aroma and taste contributions, particularly if these substances have low taste thresholds.

Allicin (**3a**) itself (and other garlic thiosulfinates, RS(O)SR', **3**) is quite reactive and fairly unstable (halflife in water at room temperature is 2-4 days), undergoing hydrolysis in water giving diallyl polysulfides (**7**, n = 2-6; Scheme 5). For optimum storage, allicin diluted with aqueous citric acid (1 mM) or water should be frozen (-20 to -70 °C). If garlic is subjected to steam distillation, a small amount of "distilled oil of garlic" is collected. This oil consists primarily of diallyl polysulfides **7** and has a garlic-like odor, somewhat different from the fresh garlic aroma characteristic of allicin (**3a**). Distilled oil of garlic is sometimes used as a substitute for fresh garlic and is the principle ingredient of one type of liquid "garlic oil" supplement. Under somewhat milder conditions, especially in the presence of cooking oils, allicin (**3a**) is transformed into ajoene **8** and dithiins **9** (Scheme 5). Ajoene and the dithiins have antithrombotic activity, and are some of the compounds responsible for the inhibition by garlic extracts of blood clot formation (*20*).

Since garlic is more often eaten cooked than raw, the effect of cooking on garlic flavorants and their decomposition products and precursors merits further discussion. All varieties of garlic have a strong aroma and taste when crushed. Boiling unpeeled whole cloves for 15 minutes, or microwaving at 650 watts for 15-30 seconds deactivates alliinase. Before this happens a small amount of allicin is formed, probably by physical contact between the cloves and each other or their container (*21*). On exposure to boiling water, allicin is converted to polysulfides **7**. If garlic is first chopped or crushed, and then heated in a closed container at 100 °C for 20 minutes, all of the initially produced allicin and other thiosulfinates are converted into polysulfides **7**. Frying garlic, which involves temperatures higher than 100 °C, converts alliin to the amino acid cysteine and allyl alcohol. Allyl alcohol most likely results from [2,3]-sigmatropic rearrangement of alliin (**1a**)(*36*). When chopped garlic is stir fried in hot oil for one minute, some polysulfides, but no allicin, remain. When a mixture of garlic, water and soybean oil is boiled, considerable allicin survives. When soybean oil is omitted, the major sulfur-containing products are polysulfides **7**. This suggests that when an edible oil is used in cooking garlic, considerable allicin may survive *moderate* heating (*21*).

Metabolism of Garlic. What happens to garlic flavorants upon ingestion? Analysis of human garlic breath by GC-AED (the subject consumed, with brief chewing, 3 g of fresh garlic with small pieces of white bread, followed by 50 mL of cold water) showed in the Se channel dimethyl selenide (MeSeMe) as the major Se component along with smaller quantities of $MeSeC_3H_5$, MeSeSMe and $MeSeSC_3H_5$; the S channel showed AllSH, MeSAll and AllSSAll with lesser amounts of MeSSMe, $MeSSC_3H_5$, an isomer of AllSSAll (presumably MeCH=CHSSAll), $C_3H_5SC_3H_5$ and $C_3H_5SSSC_3H_5$ (*32,37*). In this same study the composition of the Se and S compounds in garlic breath was examined as a function of time. After 4 hours, the levels of MeSeMe, AllSSAll, AllSAll and MeSSMe were reduced by 75% from the initial levels of 0.45 ng/L (MeSeMe), 45 ng/L (AllSSAll), 6.5 ng/mL (AllSAll), and 1.8 ng/L (MeSSMe). 2-Propenethiol (AllSH) could only be found in breath

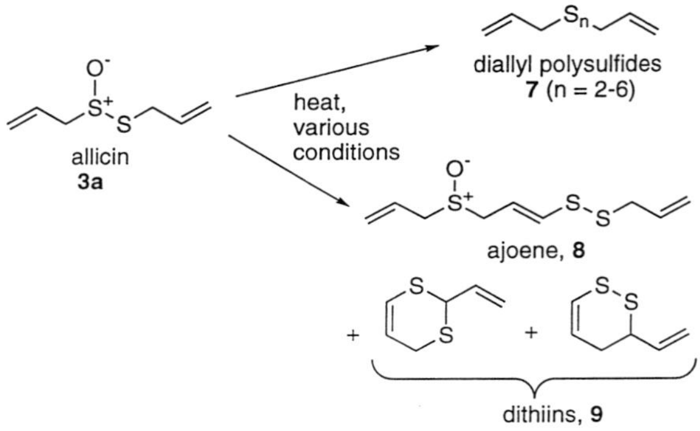

Scheme 5. Transformation of allicin into diallyl polysulfides, ajoene and dithiins.

immediately after ingestion of garlic. In view of the reported very low threshhold detection level for volatile organoselenium compounds, it is likely that MeSeMe contributes to the overall odor associated with garlic breath. In fact, MeSeMe, which has a garlic-like odor, is found in the breath air of animals fed inorganic Se compounds (*38*) and humans who have accidentally ingested Se compounds (*39*).

Consumption of larger quantities of garlic (38 g) results in persistence of levels of sulfur compounds as high as 900 ppb in the subject's breath for more than 32 hours (*40*) as well as exhalation of acetone in the breath; acetone production generally parallels increased lipid metabolism (*40*). Consumption of garlic (in the form of garlic tablets) leads to the excretion in urine of *N*-acetyl-*S*-allyl-L-cysteine (*41*). Such a metabolic product may be derived from the γ-glutamyl-*S*-allylcysteine present in fresh garlic or dried garlic powders. The rapid appearance of sulfur compounds in breath argues for rapid absorption of allicin. Indeed, administration of ^{35}S-labeled allicin to animals showed >79% absorption within 30-60 min after oral intake (*42*). Other experiments show that consumption of pure allicin lead to occurrence of substantial amounts of allyl methyl sulfide in the breath. This sulfide is most likely formed by in vivo methylation of 2-propenethiol by *S*-adenosylmethionine (*43*). The fact that allicin, ajoene, *S*-allylmercaptocysteine and diallyl trisulfide are rapidly converted to 2-propenethiol in the blood (*20*) suggests that in most cases it is not allicin itself but rather its metabolites that are the active compounds. It is well known that many drugs and nutrients require metabolic activation to convert them into bioactive forms (*44*).

Commercial Garlic Formulations

In addition to dried or pickled forms of garlic sold in food stores, garlic health supplements are available in four different forms (*43,45*): 1) Powder tablets or capsules containing dehydrated garlic powder, consisting of alliin (**1a**) and related flavor precursors, alliinase enzymes, and cysteine-containing peptides, prepared by careful dehydration of garlic slices at temperatures of 60 °C. On exposure to water this product yields up to 5 mg/g of allicin (**3a**; the "allicin-potential"). 2) Steam distilled oil of garlic, consisting of polysulfides **7**. In the People's Republic of China an antifungal agent "allitridium" is marketed, containing 14 mg/g diallyl trisulfide (*46*). 3) Garlic vegetable oil macerate extract (garlic crushed in oil), containing ajoene (**8**), dithiins (**9**) and polysulfides **7**. 4) Garlic extract aged in dilute alcohol, consisting primarily of small amounts of *S*-allylcysteine (0.6 mg/g) and cysteine-containing peptides (0.35 mg/g); no allicin is formed on exposure to water. These products are not inexpensive compared to the cost of fresh garlic, which affords the same compounds under certain conditions of food preparation and consumption. There is considerable variation in the amounts of the several garlic components in different commercial products (*21*). At the present time there are no requirements for standardization or for labeling of components or their quantities on product labels. Since crushing and processing garlic generates various pharmacologically active compounds as described above, the manufacturing process used to make a particular garlic product is a major consideration in judging its pharmacological activity. It has already been noted that the optimum pH for garlic alliinase is 6.5; the stomach pH is 1.5 when empty, 3 after a light meal, and 4.5 after a high protein meal. Alliinase is deactivated at a pH lower than 4.5, hence dehydrated garlic powder introduced into an empty stomach, or a stomach after a light meal, produces no allicin. This is why many garlic tablets are "enterically coated" so that they pass through the stomach into the less acidic small intestines before dissolving.

The Garlic Wars

In addition to debates in scholarly journals on the general efficacy of garlic preparations for treatment of various disease conditions, one manufacturer openly criticizes the basis for the purported activity of garlic cloves and of competitors' products, describing the situation as "the garlic wars." A 1997 full page ad by this manufacturer in a trade journal reads, in part: "What do you mean there's no allicin in my garlic supplement? Believe it. A prestigious scientific journal has published advanced research that conclusively proves no garlic supplement contains Allicin. Not even a trace. And that's good news. Allicin can cause oxidative damage to the cells and tissues and your body simply doesn't need it....Know your facts. If your garlic supplement is heat dried and claims to contain Allicin don't take it. Just try [Brand X]. You'll get the science of garlic without the myth of Allicin" (47). The same ad compares the S-allylcysteine (SAC) content of its product with that of six competitive powdered products and garlic oil, concluding that SAC is present in Brand X, absent in garlic oil, and of unknown concentration (indicated by "?") in the powders. However, a 1986 ad from this very same company, in criticizing a competitors claim that it "contains all the 'potential' allicin of raw garlic" states, in part: "There's nothing sadder than unrealized potential. The reality is, although that pungent sulfur compound allicin is quite volatile and difficult to measure accurately, there is actually much more allicin in [Brand X] than can be found in any other would-be competitor's product" (48). A 1991 ad by a competitor states, in part: "Modern research has learned that when fresh raw garlic is crushed, an enzyme acts on a unique amino acid to produce the reactive compound allicin. The amount of allicin produced is a test of the garlic's quality – the more allicin, the better the garlic. But allicin deteriorates within a few minutes when garlic is heated or cooked. Unlike many of the garlic oil capsules popular in America that contain no allicin, [Brand Y] has a rich allicin yield. And unlike raw garlic, [Brand Y] has none of the obvious social drawbacks of a tell-tale odor... Each [Brand Y] tablet contains all the natural constituents of 300 mg of fresh raw garlic. Unlike other garlic products, [Brand Y] maintains strict scientific standards to assure that [Brand Y] tablets contain a consistently high level of allicin" (42). The not-for-profit Center for Science in the Public Interest did an analysis of Brands X, Y and several other products (49). It found upon hydration (to generate allicin from its precursor) 4.88 mg of allicin and 0.34 mg SAC from six 100 mg tablets of Brand Y and 0 mg of allicin and 1.01 mg SAC from 4 500 mg tablets of Brand X. The per-tablet average levels of SAC found in seven other brands analyzed was 0.136 mg compared to 0.25 mg from Brand X (overall range per-tablet was from 0.045 to 0.270 mg). The allicin potential per-tablet for eight products excluding Brand X ranged from 4.8 mg to 0.114 mg. The average cost per tablet was $0.135. It was of interest that 0.33 teaspoon of a well known brand of garlic powder used for cooking on hydration produced 5.66 mg allicin and contained 0.59 mg SAC at a cost of $0.06 (*45,49*). A respected monograph on the therapeutic use of phytomedicinals (*50*) states: "most authorities now agree that the best measure of the total activity of garlic is its ability to produce allicin which, in turn results in the formation of other active principles. This ability is referred to as the allicin yield of the garlic preparation." Similarly, the British Herbal Compendium concludes that the pharmacological activity of garlic is "primarily due to allicin and/or its transformation products" (*51*).

Notes from a War Correspondent

Four strategically different camps can be identified in the "garlic wars". Those in the first camp contend, like gastroenterologist Dr. G. Woolf [quoted in *Time Magazine* (5/12/97)], that "there's no evidence that garlic does anything but make your breath smell." The second camp includes purists who argue that fresh garlic provides more substantial health benefits than commercial garlic preparations. The third and fourth camps include advocates of commercial "odorless" garlic preparations who argue respectively in support of, or against the central role of allicin (or rather "allicin potential") as represented by allicin-yielding whole garlic preparations (third camp) vs. allicin-free aged extracts of garlic (fourth camp). The author, having followed the sparring over the years, disputes the position of those in the first camp, since a substantial body of scientific evidence collected over a period of many years (epidemiology, clinical studies, in vivo-in vitro experiments, summarized below) establishes at least modest health benefits associated with regular consumption of fresh garlic.

Opposite positions in the debate on the role of allicin have been staked out (*52-54*). It should be emphasized that none of the scientifically knowledgeable manufacturers of garlic preparations contend that allicin is actually "in" their products any more than allicin is "in" fresh garlic. It should also be noted that allicin is simply the most abundant member of a family of thiosulfinates (**3**) released from garlic. The other thiosulfinates, shown in Scheme 3, should possess, and in some cases have been shown to possess, biological activity similar to that of allicin. The absence of allicin in the blood following consumption of substantial quantities of allicin-producing crushed raw garlic does not preclude the likelihood that allicin enters the blood as rapidly formed bioavailable pharmacologically active metabolites such as 2-propenethiol. It is also irrelevant whether or not allicin would be generated in the stomach since allicin would be produced in the mouth on chewing garlic or in the small intestines on consuming enterically coated tablets.

The release of allicin from enteric-coated tablets, the antibiotic effect of allicin in the intestinal tract, and the effectiveness of allicin in inhibiting cholesterol synthesis have been amply established. The beneficial physiological effects associated with compounds such as diallyl polysulfides, dithiins and ajoene, all formed by decomposition of allicin, also support the central role of allicin as either directly or indirectly being responsible for most of the health effects. Whether or not it is allicin itself, or its decomposition products or metabolites, the role of allicin as a key player in the beneficial effects of garlic seems well established. There is very limited clinical evidence to support the position that reasonable dosages of garlic-derived but allicin-free preparations have any positive health effects, apart from epidemiological studies suggesting that cooked fresh garlic (low in allicin) retains some of its anticancer effects. Indeed, no garlic-derived compounds other than thiosulfinates have ever been identified that have significant pharmacological activity at levels representative of typical garlic consumption (2-5 g/day)! Furthermore, it is ludicrous to argue (54) that an LD_{50} of allicin on the order of 60 mg/kg intravenously or 120 mg/kg subcutaneously for mice, which translates to 1,000 to 2,000 g of raw garlic by a person in a single dose (intravenously or subcutaneously!), establishes the toxicity of allicin in garlic under conditions of typical mealtime use, where for example 12 mg of allicin is released by each 3 g clove of garlic consumed.

Erroneous calculations have been employed by the anti-allicin camp to present allicin as possibly being harmful. Thus, it is stated (52) that "One milliliter of an allicin solution (53 mg/mL in EtOH) was pipetted into a 50 mL volumetric flask, and saline was added to volume. Heparinized blood (1 mL) was pipetted into a 50 mL volumetric flask, and saline was added to volume. Five milliliters of this solution were transferred to a 25 mL volumetric flask, 5 mL of the diluted allicin solution was added, and saline was added to volume. The visible spectrum ... was obtained". Since both the 53 mg of allicin and the 1 mL of blood underwent *identical dilutions*, the final concentration remains "53 mg allicin/mL of blood" representing "25,000 times the maximum possible dose achievable by consumption of a typical 3 g clove. To achieve this dose, a person would have to eat his/her own body weight in fresh garlic in 1 min." (53). The fact that the final concentration of allicin after the above dilution if 0.212 mg/mL (54) is irrelevant since the same 250 fold dilution was applied to the original 1 mL of blood, maintaining the ratio at "53 mg allicin/mL of blood." Indeed, almost any vitamin, water, salt, etc. can be harmful or cause death if taken all at once in large enough quantities.

Health Effects of Garlic

In recent years, fresh garlic and garlic preparations have been shown to lower serum cholesterol and triglycerides levels and blood pressure, to possess antithrombotic, antibiotic, anticancer, and antioxidant activity, and to stimulate the immune system (10,20,21,43,45, 55). The 1992 *British Herbal Compendium* (51) attributes the following types of therapeutic activity to garlic: "lowers blood cholesterol and triglycerides, hypotensive, lowers blood viscosity, activates fibrinolysis, inhibits platelet aggregation; antimicrobial, anti-inflammatory, anthelmintic." The 1988 German "Kommission E" monograph has approved use of garlic products to supplement dietary measures in cases of elevated blood lipid levels [hyperlipoproteinemia] and for the prophylaxis of age-related vascular changes [atheriosclerosis] (50,51). It also attributes antibacterial and antimycotic activity to garlic preparations. In the People's Republic of China, the drug "allitridium", which is mainly diallyl trisulfide, is employed together with amphotericin B as an antifungal drug (*46*). The activity of allitridium against *Cryptococcus neoformans* has been confirmed through in vitro studies (*46*). Brief mention is made of recent in vivo and in vitro research on the various health effects of garlic. Since garlic thiosulfinates (as well as ajoene) react readily with thiols such as cysteine and undergo hydrolysis, the *in vivo* lifetimes of these compounds are quite limited. Metabolites of these compounds are therefore more likely to be the active agents *in vivo*.

What's in a Name? It has been noted above that since crushing and processing garlic generates various pharmacologically active compounds, the process used to make a particular garlic product is a major consideration in judging its pharmacological activity. Unfortunately, many papers describing health effects of materials derived directly or indirectly from garlic simply refer to the material as "garlic" or "*Allium sativum*" even though the chemical composition may be vastly different from that of the fresh, intact herb. In some cases the precise chemical composition of the material used, or details of its preparation, is not even specified. This lack of compositional specificity renders many published studies of physiological activity of garlic-derived products worthless! The discussion of health effects that follows is limited to those cases where the nature of the garlic preparation used is clearly indicated.

Antibiotic Effects. Numerous clinical and laboratory studies have established the antibacterial and antifungal activity of crushed garlic and have demonstrated the effectiveness of garlic extracts against a wide range of gram-positive and gram-negative bacteria (*21, 45*). At low concentrations, garlic extracts display antimicrobial activity against *Helicobacter pylori*, the cause of gastric ulcers. This effect may be related to the lower risk of stomach cancer in those with a high Allium vegetable intake (*56*). While allicin and related thiosulfinates have been shown to be the principle antibiotics in garlic extracts (*20,21,57*), both ajoene and diallyl trisulfide have been shown to inhibit the growth of some pathogenic fungi (*46*). Because both allicin and ajoene have very short lifetimes in blood and other bodily fluids, the direct antibiotic activity of these garlic-derived compounds is largely limited to topical usage or to the intestines, stomach and esophagus, if in direct contact with freshly formed allicin or ajoene. However, systemic antimicrobial effects have been reported both in animals and humans for garlic preparations, suggesting that allicin and ajoene metabolites are also active (*43*). Allicin has been reported to inhibit the cytopathic effects of the ameoba *Entamoeba histolytica* (*58*), which is consistent with traditional usage of garlic to treat ameobic dysentary (*19*).

Anticancer Effects. It has been known for some time that cancer occurs least in those countries where garlic is eaten regularly (*21*). Epidemiological studies show a consistent correlation between consumption of garlic and decreased risk of gastrointestinal-colorectal cancers (*21,59*). A recent study suggests that increased intake of garlic is also associated with decreased risk of prostate cancer (*60*). A 1996 study of occurrence of adenomatous colorectal polyps (considered precursors to colorectal cancers) found the strongest inverse association with increased consumption of carotenoid-rich vegetables, cruciferae (including broccoli), and garlic (*61*). Ajoene (**8**) has been identified as one of the effective antimutagenic compounds derived from garlic (*62*) while *S*-allylmercaptocysteine, formed on reaction of allicin or ajoene with cysteine, inhibits the proliferation of various cancer cell lines. Both allicin and ajoene have been found to be cytotoxic to three different fast-growing cell lines, with EC$_{50}$ values in the micromolar range (*63*), but less so by two to three orders of magnitude to various liver cells (*64*). The low cytotoxicity of both compounds in liver cells is said to be in accord with the observation that garlic consumption even in high doses is well tolerated (*65*).

Aqueous extracts of garlic as well as diallyl sulfide and *S*-allylcysteine prevent benzo(a)pyrene-DNA adduct formation in stimulated human peripheral blood lymphocytes in vitro and therefore may be useful in preventing benzo(a)pyrene associated tumorigenesis (*65*). Diallyl disulfide has been found to inhibit the proliferation of human tumor cells in culture (*66*). It has been suggested that trisulfides and other polysulfides, in concert with thiols, can mediate the reduction of molecular oxygen to DNA-cleaving oxygen radicals and thereby act as antitumor agents (*67*). Antibacterial properties of garlic may inhibit bacterial conversion of nitrate to nitrite in the stomach, thereby limiting formation of carcinogenic nitrosamines (*68*). Alternatively, garlic sulfur compounds may inhibit *N*-nitrosation by rapidly forming *S*-nitrosothiols or thionitrites and nitric oxide; the *S*-nitroso compounds would then decompose giving disulfides (*69*). 2-Propenethiol (allyl mercaptan) is particularly effective in inhibiting *N*-nitrosodiethylamine-induced forestomach tumor formation (*21*). This is significant because allicin and diallyl trisulfide are known to be rapidly converted to 2-propenethiol in the blood (*21*), as demonstrated by immediate detection of 2-propenethiol in breath after consumption of fresh garlic (*32,37*). The anticancer activity of garlic sulfur compounds has also been associated with induction of cellular glutathione, a major intracellular antioxidant, and stimulation of glutathione *S*-transferase (*21*), an enzyme which can detoxify and dispose of harmful chemicals, inhibition of ornithine decarboxylase, an enzyme important in DNA synthesis, whose unregulated activity occurs in neoplasms, or by other activity of these compounds as antioxidants or nucleophiles to provide defense against peroxides and alkylating agents.

Antioxidant Effects. Dietary antioxidants are of value in reducing the risk of atherosclerosis and cancer. Garlic, allicin and diallyl polysulfides all possess antioxidant activity as established by decreased lipid peroxidation, and increased free radical scavenging (*21*). Of 22 common vegetables, garlic shows the highest antioxidant activity toward the peroxyl radical (*70*). In clinical trials, ingestion of garlic powder tablets for two months led to a significant reduction in levels of serum malondialdehyde, a marker for lipid peroxidation, as well as an increase in levels of erythrocyte reduced glutathione (*71*). Purified alliin (**1a**) has no antioxidative activity in the linoleic acid oxidation system (*72*). The in vitro antioxidant activity of dialkyl thiosulfinates **3** has been known since 1961 (*73-77*) and is associated in part with formation of sulfenic acids **2**, which are extremely active radical scavengers (*78*). The lipoxygenase inhibitory activity of garlic distilled oil components has also been reported (*79*). Allicin is reported to oxidize hemoglobin to methemoglobin (*52*), but only at a level representing 25,000 times the maximum possible dose achievable by consumption of a typical 3 g clove (*53*). 1-Propenyl compounds, found in garlic and onion distilled oils, can in higher dosages promote some types of tumor formation by serving as pro-oxidants. Thiyl radicals, e.g. those formed from Allium-derived disulfides and thiols by one-electron oxidation, can also act as pro-oxidants (*55*). Thus, it may be more appropriate to view Allium-derived sulfur compounds as redox agents, e.g. antioxidant in some circumstances, pro-oxidant in others (*55*).

Antithrombotic and Related Activity. In healthy blood function, blood coagulation is physiologically balanced by fibrinolysis (enzymatic splitting of the fiberous protein fibrin). Impairment of the fibrinolytic activity enhances blood clotting and thrombosis, with the attendant risk of myocardial and cerebral infarction. Platelet aggregation also plays a key role in coagulation and in hemostasis (stoppage of bleeding). Garlic juice or oils enhances fibrinolysis, based on both clinical and epidemiological studies, while garlic inhibits platelet aggregation in animals and in humans (*21*). Garlic-derived platelet inhibitors include allicin, ajoene, dithiins, and diallyl polysulfides; *S*-allylcysteine has no activity. Most of the human clinical studies suggest that consumption of moderate amounts of fresh garlic and whole clove commercial garlic products has a favorable effect on fibrinolysis (*80*), and in some cases can inhibit spontaneous platelet aggregation while also increasing capillary circulation. Administration of garlic macerate capsules (rich in dithiins and ajoene) for one month to healthy subjects and patients with coronary artery disease led to significant ex vivo inhibition of platelet aggregation (*81*). Rats given a low dose of an aqueous extract of garlic exhibit significant reduction in levels of serum thromboxane B2 (*82*). Because of the inhibiting effect of garlic on blood clotting, surgeons have begun asking patients to abstain from garlic prior to surgery (*83,84*).

Cholesterol and Triglyceride Lowering Effects. Allicin displays lipid biosynthesis inhibitory activity (*21,45*). The inhibitory potential toward cholesterol biosynthesis as determined using rat hepatocyte cultures decreases in the order allicin > diallyl disulfide > 2-propenethiol = 2-vinyl-4*H*-1,3-dithiin >> 3-vinyl-4*H*-1,2-dithiin (*85*). The fact that allicin was the most effective inhibitor and that 2-propenethiol showed little activity argues against the possibility that allicin degradation products (diallyl disulfide or 2-propenethiol) might mediate its activity. Clinical studies involving coadministration of garlic powder pills and fish oil (rich in n-3 fatty acids) lowered total cholesterol, LDL-C, and triglyceride concentrations (*86*).

Blood-Pressure Lowering Effects. The allicin precursor γ-glutamyl *S*-allyl cysteine has been found to inhibit the blood pressure-regulating angiotensin converting en-zymes (*45*). Garlic extracts have been found to increase NO synthase activity supporting the claim that garlic exerts some of its therapeutic properties, e.g. its ability to decrease blood pressure, by increasing NO production in the body (*87-89*).

140

Too Much of a Good Thing? While aspirin at dosages of up to one tablet per day is recommended for prevention of cardiovascular diseases, among other conditions, larger daily dosages can cause gastric ulceration and internal bleeding. For similar reasons, ingestion of "excessive" amounts of raw garlic can also be injurious and should be avoided (*21*). While the definition of "excessive" varies from individual to individual and depends on whether the raw garlic is taken with food or on an empty stomach, an example would be the individual who consumed "five to ten raw cloves ... daily over ten years" and developed hemorrhagic gastritis (*90*).

Acknowledgments

I thank my skilled colleagues, whose names are listed in the references, for contributing to the work reported herein, and Debra Jahner and Larry Lawson for providing recent references and for helpful suggestions. This work was supported by grants from NSF, NIH-NCI, the NRI Competitive Grants Program/USDA (Awards No. 92-37500-8068 and 96-35500-3351), and McCormick & Company.

Literature Cited

1. Blumenthal, M. *HerbalGram* **1995**, *34*, 36.
2. Brevoort, P. *HerbalGram* **1996**, *36*, 49.
3. Gruenwald, J. *HerbalGram* **1995**, *34*, 60.
4. Dreidger, S.D. *Maclean's*, May 20, 1996, *109*, 62.
5. Anonymous, *National Examiner*, October 30, 1984, 1.
6. Anonymous, *Globe*, February 14, 1989, 1.
7. Anonymous, *Weekly World News*, December 13, 1988, 1.
8. Eftekhar, J.L. *Garlic -- the Miracle Herb*, Globe Communications: Boca Raton, FL; 1995.
9. Moffett, M. *Garlic, Selenium, Cranberries & Green Tea*, MicroMags: Lantana, FL; 1997.
10. Tyler, V.E. *The Honest Herbal*; Third Edition; Pharmaceutical Products Press [Haworth Press Imprint]: New York, 1993.
11. Beaglehole, R. *Lancet* **1996**, *348*, 1186.
12. Ernst, E. *Lancet* **1997**, *349*, 131.
13. Rahman, K.; Billington, D.; Rigby, G. *Lancet* **1997**, *349*, 131.
14. Myers, S.P.; Smith, A.J. *Lancet* **1997**, *349*, 131.
15. Beaglehole, R. *Lancet* **1997**, *349*, 131.
16. Neil, H.A.W.; Silagy, C.A.; Lancaster, T.; Hodgeman, J.; Vos, K.; Moore, J.W.; Jones, L.; Cahill, J.; Fowler, G.H. *J. Roy. Coll. Physicians* **1996**, *30*, 329.
17. Simons, L.A.; Balasubramaniam, S.; von Konigsmark, M. *Atherosclerosis* **1995**, *113*, 219.
18. Jain, A.K.; Vargas, R.; Gotzkowsky, S.; McMahon, F.G. *Am. J. Med.* **1993**, *94*, 632.
19. Block, E. *Sci. Amer.* **1985**, *252*, 114.
20. Block, E. *Angew. Chem., Int. Ed. Engl.* **1992**, 31, 1135.

21. Koch, H.P., Lawson, L.D., (Ed.) *Garlic. The Science and Therapeutic Application of Allium sativum L. and Related Species*. 2nd Ed. Williams & Wilkins: Baltimore, MD; 1996 [out of print: order from American Botanical Council, Austin, TX].

22. Block, E. In *1998 Medical and Health Annual*, Encyclopaedia Britannica, Chicago, 1997, pp 222-229.

23. Calvey, E.M.; Block, E. In *Spices: Flavor Chemistry and Antioxidant Properties*, ACS Symposium Series 660, S. J. Risch, C.-T. Ho, Ed., American Chemical Society: Washington DC; pp 113-124, **1997**.

24. Calvey, E.M.; Matusik, J.E.; White, K.D.; DeOrazio, R.; Sha, D.; Block, E. *J. Agric. Food Chem.* **1997**, *45*, 4406.

25. Lawson, L.D.; Wang, Z.Y.J.; Hughes, B.G. *J. Nat. Prod.* **1991**, *54*, 436.

26. Ip, C.; Lisk, D.J.; Stoewsand, G.S. *Nutr. Cancer* **1992**, *17*, 279.

27. Ip, C.; Lisk, D.J. *Carcinogenesis* **1994**, *15*, 1881.

28. Lu, J.; Pei, H.; Ip, C.; Lisk, D.J.; Ganther, H.; Thompson, H.J. *Carcinogenesis* **1996**, *17*, 1903.

29. Ip, C.; Lisk, D.J.; Thompson, H.J. *Carcinogenesis* **1996**, *17*, 1979.

30. Ip, C.; Lisk, D.J. *Adv. Exp. Med. Biol.* **1996**, *401*, 179.

31. Clark, L.C.; Combs, G.F.; Turnball, B.W.; Slate, E.H.; Chalker, D.K.; Chow, J.; Davis, L.S.; Glover, R.A.; Graham, D.K.; Gross, E.G.; Krongrad, A.; Lesher, J.L.; Park, H.K.; Sanders, B.B.; Smith, C.L.; Taylor, J.R.; Alberts, D.S.; Allison, R.J.; Bradshaw, J.C.; Curtis, D.; Deal, D.R.; Dellasega, M.; Hendrix, J.D.; Herlong, J.H.; Hixson, L.J.; Knight, F.; Moore, J.; Rice, J.S.; Rogers, A.I.; Schuman, B.; Smith, E.H.; Woodard, J.C. *J. Am. Med. Assoc.* **1996**, *276*, 1957.

32. Block, E.; Cai, X.-J.; Uden, P.C.; Zhang, X.; Quimby, B.D.; Sullivan, J.J. *Pure Appl. Chem.* **1996**, *68*, 937.

33. Cai, X.-J.; Uden, P.C.; Block, E.; Zhang, X.; Quimby, B.D.; Sullivan, J.J. *J. Agric. Food Chem.* **1994**, *42*, 2081.

34. Cai, X.-J.; Block, E.; Uden, P.C.; Zhang, X.; Quimby, B.D.; Sullivan, J.J. *J. Agric. Food Chem.* **1995**, *43*, 1754.

35. Bird, S.M.; Ge, H.; Uden, P.C.; Tyson, J.F.; Block, E.; Denoyer, E. *J. Chromatog. A* **1997**, *789*, 349.

36. Yu, T.-H.; Wu, C.-M.; Rosen, R. T.; Hartman, T. G.; Ho, C.-T. *J. Agric. Food Chem.* **1994**, *42*, 146.

37. Cai, X.-J.; Block, E.; Uden, P.C.; Quimby, B.D.; Sullivan, J.J. *J. Agric. Food Chem.* **1995**, *43*, 1751.

38. Oyamada, N.; Kikuchi, M.; Ishizaki, M. *Anal. Sci.* **1987**, *3*, 373.

39. Buchan, R.F. *J. Am. Med. Assoc.* **1974**, *227*, 559.

40. Taucher, J.; Hansel, A.; Jordan, A.; Lindinger, W. *J. Agric. Food Chem.*, **1996**, *44*, 3778.

41. de Rooij, B.M.; Boogaard, P.J.; Rijksen, D.A.; Commandeur, J.N.; Vermeulen, N.P. *Arch. Toxicol.* **1996**, *70*, 635.

42. Lachmann, G.; Lorenz, D.; Radeck, W.; Steiper, M. *Arzneim. Forsch.* **1994**, *44*, 734.

142

43. Lawson, L.D. In *Phytomedicines of Europe: Their Chemistry and Biological Activity*, ACS Symposium Series 691, L.D. Lawson, R. Bauer, Ed., American Chemical Society: Washington DC; pp 175-208, **1998**.

44. Lawson, L.D.; Wang, Z.Y.J. *Planta Medica* **1993**, *59*, A688.

45. Block, E. In *Hypernutritious Foods*, Finley, J.W.; Armstrong, D.J.; Nagy, S.; Robinson, S.F. Ed. Agscience: Auburndale FL, 1996, Ch. 16.

46. Shen, J.; Davis, L.E.; Wallace, J.M.; Cai, Y.; Lawson, L.D. *Planta Medica* **1996**, *62*, 415-418.

47. *Vegetarian Times*, August 1997, 13.

48. Copies of advertisements are available from the author upon written request.

49. Schardt, D.; Liebman, B. *Nutr. Action Health Lett.* **1995**, *22* (6), 3.

50. Tyler, V.E. *Herbs of Choice*. Pharmaceutical Products Press [Haworth Press Imprint]: New York, 1994.

51. Bradley, P.R., Ed. *British Herbal Compendium*. British Herbal Medicine Association: London, 1992; vol. 1, "Garlic", p.105.

52. Freeman, F.; Kodera, Y. *J. Agric. Food Chem.* **1995**, *43*, 2332-2338.

53. Lawson, L.D.; Block, E. *J. Agric. Food Chem.* **1997**, *45*, 542.

54. Freeman, F.; Kodera, Y. *J. Agric. Food Chem.* **1997**, *45*, 3709.

55. Block, E. *Adv. Exp. Med. Biol.* **1996**, *401*, 155.

56. Sivam, G.P.; Lampe, J.W.; Ulness, B.; Swanzy, S.R.; Potter, J.D. *Nutrition & Cancer* **1997**, *27*, 118.

57. Cavallito, C.J.; Bailey, J.H. *J. Am. Chem. Soc.* **1944**, *66*, 1950.

58. Ankri, S.; Miron, T.; Rabinkov, A.; Wilchek, M.; Mirelman, D. *Antimicrob. Agents Chemother.* **1997**, *41*, 2286.

59. Steinmetz, K.A.; Kushi, L.H.; Bostick, R.M.; Folsom, A.R.; Potter, J.D. *Am. J. Epidemiol.* **1994**, *139*, 1.

60. Key, T.J.; Silcocks, P.B.; Davey, G.K.; Appleby, P.N.; Bishop, D.T. *Br. J. Cancer* **1997**, *76*, 678.

61. Witte, J.S.; Longnecker, M.P.; Bird, C.L.; Lee, E.R.; Frankl. H.D.; Haile, R.W. *Amer. J. Epidemiol.* **1996**, *144*, 1015.

62. Ishikawa, K.; Naganawa, R.; Yoshida, H.; Iwata, N.; Fukuda, H.; Fujino, T. Suzuki, A. *Biosci. Biotech. Biochem.* **1996**, *60*, 2086.

63. Scharfenberg, K.; Wagner, R.; Wagner, K.G. *Cancer Letters* **1990**, *53*, 103.

64. Gebhardt, R.; Beck, H.; Wagner, K.G. *Biochim. Biophys. Acta* **1994**, *1213*, 57.

65. Hageman, G.J.; van Herwijnen, M.H.M.; Schilderman, P.A.E.L; Rhijnsburger, E.H.; Moonen, E.J.C.; Kleinjans, J.C.S. *Nutrition & Cancer* **1997**, *27*, 177.

66. Sundaram, S.G.; Milner, J.A. *Biochim. Biophys. Acta* **1996**, *1315*, 15.

67. Mitra, K.; Kim, W.; Daniels, J.S.; Gates, K.S. *J. Am. Chem. Soc.* **1997**, *119*, 11691.

68. Mei, X.; Wang, M.C.; Xu, H.X.; Pan, X.P.; Gao, C.Y.; Han, N.; Fu, M.Y. *Acta Nutrimenta Sinica* **1982**, *4*, 53.

69. Shenoy, N.R.; Choughuley, A.S.U. *Cancer Lett.* **1992**, *65*, 227.

70. Cao, G.; Sofic, E.; Prior, R.L. *J. Agric. Food Chem.* **1996**, *44*, 3426.

71. Grune, T.; Scherat, T.; Behrend, H.; Conradi, E.; Brenke, R.; Siems, W. *Phytomed.* **1996**, *2*, 205.

72. Hirata, R.; Matsushita, S. *Biosci. Biotech. Biochem.* **1996**, *60*, 484.
73. Barnard, D.; Bateman, L.; Cain, M.E.; Colclough, T.; Cunneen, J.I. *J. Chem. Soc.* **1961**, 5339.
74. Bateman, L.; Cain, M.E.; Colclough, T.; Cunneen, J.I. *J. Chem. Soc.* **1962**, 3570.
75. Rahman, A.; Williams, A. *J. Chem. Soc. B*, **1970**, 1391.
76. Block, E. *J. Am. Chem. Soc.* **1972**, *94*, 644.
77. Block, E.; O'Connor, J. *J. Am. Chem. Soc.* **1974**, *96*, 3929.
78. Koelewijn, P.; Berger, H. *Recl. Trav. Chim. Pays-Bas* **1972**, *91*, 1275.
79. Block, E.; Iyer, R.; Grisoni, S.; Saha, C.; Belman, S.; Lossing, F.P. *J. Am. Chem. Soc.* **1988**, *110*, 7813.
80. Gong, X.; Ouyan, Z.; Cai, D. *Tianran Chanwu Yanjiu Yu Kaifa* **1996**, *8*, 59-62.
81. Bordia, A.; Verma, S.K.; Srivastava, K.C. *Prostaglandins, Leukot. Essent. Fatty Acids* **1996**, *55*, 201.
82. Bordia, T.; Mohammed, N.; Thomson, M.; Ali, M. *Prostaglandins, Leukot. Essent. Fatty Acids* **1996**, *54*, 183.
83. Burnham, B.E. *Plast. Recon. Surg.* **1995**, *95*, 213.
84. Petry, J.J. *Plast. Recon. Surg.* **1995**, *96*, 483.
85. Gebhardt, R.; Beck, H. *Lipids* **1996**, *31*, 1269.
86. Adler, A.J.; Holub, B.J. *Am. J. Clin. Nutr.* **1997**, *65*, 445.
87. Das, I.; Khan, N.; Sooranna, S.R. *Curr. Med. Res. Opin.* **1995**, *13*, 257.
88. Das, I.; Khan, N.; Sooranna, S.R. *Biochem. Soc. Trans.* **1995**, *23*, 136S.
89. Sooranna, S.R.; Hirani, J.; Das, I. *Biochem. Soc. Trans.* **1995**, *23*, 543S.
90. Letter to the Editor, *Alternative Medicine Digest*, December 1997, Issue 21, p 6.

Chapter 14

Garlic and Associated Allyl Sulfur Constituents Depress Chemical Carcinogenesis

J. A. Milner and E. M. Schaffer

Nutrition Department and Graduate Program in Nutrition, The Pennsylvania State University, University Park, PA 16802

Increased intake of garlic and related *Allium* foods is proposed to offer protection against cancer. The ability of water-extracts of garlic and onion, but not leeks, to block spontaneous nitrosamine formation suggests all *Allium* foods are not equal in their anticarcinogenic properties. Likewise, the depression in nitrosamine formation resulting from S-allyl cysteine (SAC) treatment, but not diallyl disulfide (DADS), suggests not all allyl sulfur compounds can be assumed equal. Nevertheless, SAC and DADS are both effective in blocking nitrosamine bioactivation. The ability of dietary garlic powder supplements to inhibit 7,12 dimethylbenz(a)anthracene (DMBA) induced mammary carcinogenesis and DNA adducts suggest this food has widespread anticarcinogenic properties. The composition of the diet can markedly modify garlic's ability to reduce chemical carcinogenesis. Alterations in phase I and II enzymes, including prostaglandin synthase, may explain some of the anticarcinogenic properties of garlic. Collectively, the ability of garlic and associated allyl sulfur compounds to depress carcinogen formation and bioactivation is consistent with epidemiological evidence indicating that higher intake of *Allium* foods is accompanied by a reduction in the risks of some cancers.

Cancer, once thought an inevitable consequence of aging, is now considered by many to reflect cellular reactions to a myriad of environmental factors, including dietary habits. The connection between cancer risk and dietary habits is unquestionably complex and reflects dynamic interactions occurring among essential and non-essential dietary components. Despite the plethora of both positive and negative nutrient interactions, it is estimated that approximately 30 to 60% of all cancers relate to dietary habits (*1*). Not surprisingly, the intake of some foods correlates with an increase in risk, while others appear to provide some degree of protection. One of the food groups whose intake is associated with reduced cancer risk comes from the genus *Allium* (*2,3*). While, these foods (which include garlic, onions, leeks, and chives) have been used for centuries for their antimicrobial properties, their ability to repel insects, and their characteristic aroma and flavor, they have only recently received significant attention as naturally occurring sources of anticarcinogenic and antitumorigenic agents. This review will focus on some of the more recent epidemiological and laboratory findings about the anticarcinogenic properties of garlic and related foods.

Epidemiological Evidence that Allium Foods Inhibit Cancer

Epidemiological evidence, although limited, provides clues about the potential health benefits associated with increased *Allium* food consumption. Several studies have observed that increased garlic consumption correlates inversely with cancer mortality (*4-8*). Mei et al. (*8*) reported that individuals consuming 20 g garlic per d had about 10% of the risk of gastric cancer mortality of those consuming 2 g per d. Recent information suggests the anticancer benefits of garlic may accrue at substantially lower intakes (*6*). In the study of Steinmetz et al. (*6*) the intake of garlic, about 0.6 g per week, was the only food item among 127 evaluated that correlated with a reduction in colon cancer risk in the almost 42,000 women examined. It remains to be determine why some epidemiological studies, but not all (*9-11*), provide evidence that garlic and related foods have anticarcinogenic properties. Some of the discrepancy among studies may relate to the interaction of garlic with other dietary components. Amagase et al. (*12*) observed that the dietary intake of protein, methionine, lipid, vitamin A and selenium could significantly modify the ability of garlic to reduce the bioactivation of the mammary carcinogen 7, 12 dimethylbenz(a)anthracene (DMBA). Whether or not these are other dietary interactions account for some of the variability in cancer risk in humans remains to be determined.

Garlic and Nitrosamine Biosynthesis

Garlic has long been touted for its antimicrobial properties. Water-extracts of garlic are recognized to possess broad-spectrum activity against bacteria and fungi (*13*). This antimicrobial property may account for the reduced bacterial and fungal-mediated synthesis of nitrite and ultimately the formation of carcinogenic nitrosamines in studies by Mei et al (*14*). However, it appears that the ability of garlic to reduce nitrosamine formation may occur independent of its antimicrobial properties. Dion et al. (*15*) found that a water-extract of several sources of garlic suppressed the spontaneous (chemical) formation of nitrosamines (Figure 1). Since most nitrosamines are recognized for their carcinogenic potential, this effect alone may significantly influence human cancer risk. While the mechanism by which a water-extract of garlic leads to this effect remains largely unexplored, it is possible as suggested by Shenoy and Choughuley (*16*) that it relates to the formation of S-nitrosothiols.

Various *Allium* foods have the propensity to suppress spontaneous nitrosamine formation (*15*). However, not all are equivalent since garlic and onions, but not leeks, were effective in blocking the spontaneous formation of nitrosomorpholine from its precursors (Figure 1). S-allyl cysteine (SAC) a water-soluble sulfur compound but not diallyl disulfide (DADS), a lipid-soluble allyl sulfur compound, was found to suppress spontaneous nitrosomorpholine (NMOR) formation (*15*). Both SAC and DADS were effective in reducing the formation of revertants in a liver microsomal activated *Salmonella typhimurium* TA100 assay to which NMOR was added. While the addition of 7 μmoles SAC per plate reduced NMOR mutagenicity by 22%, adding as little as 0.03 μmoles DADS was accompanied by a 69% decrease in its mutagenicity. While DADS was considerably more effective than SAC in blocking the mutagenicity of NMOR, both were more effective than isomolar ascorbic acid (15). Collectively, these investigations demonstrate that garlic and related sulfur compounds can inhibit carcinogenesis by at least two mechanisms, namely suppressing carcinogen formation and inhibiting carcinogen bioactivation.

The abilty of a water extract of garlic to suppress the mutagenicity of nitrosamines is consistent with previous observations that dietary garlic supplementation retards liver DNA adducts caused by treatment with preformed nitrosamines (17). This suppression in adducts likely reflects a suppression in carcinogen bioactivation possibly attributable to altered cytochrome p450 activity. Brady et al. (18) found that diallyl sulfide (DAS), another oil-soluble allyl compound from processed garlic, could inactivate cytochrome p450 2E1 (CYP2E1). The metabolism of chlorzoxazone has been used as an *in vivo* biomarker for CYP2E1 activity (19). Recent studies from our laboratory reveal that dietary intake of garlic (2%), SAC (57 mmol/kg) and DADS (57 mmol/kg) decrease the hydroxylation of chlorzoxazone (20). A depression in cytochrome p450 may explain the ability of DAS to completely inhibit N-nitrosomethylbenzylamine-induced esophageal tumors (21). Increased glutathione and glutathione-S-transferases (factors involved in phase II detoxification) also has been observed following the consumption of processed garlic powder in experimental animals (22). While the response in phase II enzymes and substrates likely is highly dependent on the quantity of garlic or associated constituents consumed, these changes may contribute significantly to the anticarcinogenic properties of garlic.

Garlic Blocks Action of Other Carcinogens

The protection offered by garlic against chemically induced tumors is not limited to nitrosamines. Several studies have shown that garlic and its associated sulfur compounds are effective in blocking the incidence of tumors in experimental animals treated with polycyclic hydrocarbons, such as 7,12 dimethylbenz(a)anthracene (DMBA) (23), and alkylating compounds, such as 1,2 dimethylhydrazine (24). Topical application of garlic oil was shown several years ago to substantially reduce skin cancers DMBA treated mice (25). Studies from our laboratory show that the occurrence of DMBA induced DNA adducts in mammary tissue 24 h after carcinogen treatment is highly correlated with ultimate tumor incidence (23). Thus, alterations in these DNA adducts can serve as a sensitive biomarker of the impact of dietary constituents on DMBA carcinogenesis. Studies from our laboratory reveal that both SAC and DADS markedly inhibit the occurrence of DNA adducts and the carcinogenicity of DMBA (23, 26, 27). This protection results from biological effects occurring during both the initiation and promotion phases of carcinogenesis (23).

Garlic preparations likely vary in the amount of specific organosulfur compounds. Part of this variation may be attributable to the growing conditions. Thus, not all garlic sources can be assumed to be equally effective blockers of chemically induced tumors in experimental animals, or presumably in humans. The quantity of SAC occurring in various garlic preparations was found to correlate with their ability to inhibit the *in vivo* formation of DMBA-DNA adducts (26). Nevertheless, in addition to allyl sulfur compounds constituents such as arginine, polysaccharides, saponins and/or selenium likely influence the anticarcinogenic potential of garlic. Studies by Ip et al. (28) reveal that variation in the content of selenium in garlic can markedly influence its ability to influence DMBA carcinogenesis. While additional attention is needed to evaluate the influence of processing on the ability of garlic and related constituents to inhibit the cancer process, studies from our laboratory suggest that crushed and dehydrated garlic, commercially processed high sulfur garlic and commercially processed deodorized garlic are all effective in reducing the binding of DMBA to mammary cell DNA (26). Clearly, our studies demonstrate that odor is not a necessary prerequisite for the ability of garlic preparations to inhibit chemically induced DNA adducts or tumors (23, 26, 27).

As indicated previously several components of the diet can influence the protection provided by garlic against the cancer process. Schaffer et al. (27) and Ip et al. (28) have shown that providing selenium either as a constituent of the garlic or as a separate supplement can enhance the protection provided by garlic. It remains to be determine what metabolic changes account for this combined benefit. Studies from our laboratory show that glutathione S-transferase and UDP-glucuronosyltransferase activities were enhanced by dietary selenite supplementation, while the garlic, SAC or DADS do not significantly affect activities of these enzymes (27). SAC and DADS may be modifying phase I and II enzymes other than those influenced by selenium intake.

Our laboratory has also shown that the quantity and type of lipid consumed by rats can also influence the ability of garlic to depress DNA adducts in mammary tissue resulting from treatment with DMBA (Figure 2) (12). Inclusion of 2% deodorized garlic in the diet inhibited DMBA-induced mammary DNA adducts by approximately 50% when the diet contains 20% corn oil, but has virtually no effect when the diet contains 5% corn oil (12). The impact of lipid intake likely relates to the intake of specific fatty acids. Recent studies from our laboratory have shown that while dietary linoleic acid and oleic acid supplementation can markedly enhance the presence of mammary cell DNA adducts following treatment with DMBA, a similar effect is not seen when palmitic acid is provided (29). Inclusion of garlic in the diet has the greatest inhibitory effect on DMBA carcinogenesis when the diet contains high linoleic acid than high amounts of oleic acid (30) (Figure 3).

Bioactivation of benzo[a]pyrene, another polycyclic aromatic hydrocarbon, is known to be mediated by prostaglandin synthase (PHS) (31). We undertook studies to determine if PHS completes DMBA bioactivation and if the garlic constituents SAC and/or DADS might alter this reaction (Figure 4). Indeed, PHS was found to mediate the formation of DNA adducts when 3,4-DMBA-dihydrodiol was incubated with PHS, arachidonic acid and calf-thymus DNA (Figure 5). Indomethacin is a recognized inhibitor of PHS activity (32). Interestingly, the addition of DADS or SAC to the incubation media in the present studies was also found to significantly inhibit the PHS mediated formation of DNA adducts in vitro (Figure 5). DADS and SAC were approximately 20 and 45% less effective on a molar basis than was indomethacin in suppressing the bioactivation of 3,4-DMBA-dihydrodiol. The present studies suggest PHS activity may represent a key site of action of garlic and related sulfur compounds in blocking DMBA carcinogenesis.

At least one other mechanism may account for the anticarcinogenic properties of garlic and related allyl sulfur compounds. Supplemental dietary garlic has been found to significantly reduce adducts caused by the treatment with methylnitrosurea, a carcinogen that does not require metabolic activation. These results raise the possibility that enhanced DNA repair may also be occurring following increased garlic consumption (17). More recent studies have demonstrated that feeding garlic powder (2%), SAC (57 mmol/kg) or DADS (57 mmol/kg) are all effective in decreasing the incidence of mammary tumors resulting from treatment with this carcinogen (76, 41, and 53% inhibition, respectively) in a 23 week feeding study (33).

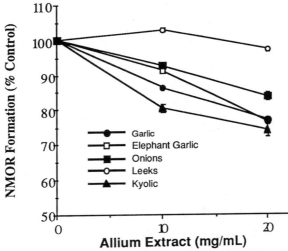

Figure 1. Effect of water extracts of garlic, elephant garlic, leeks, onions and deodorized garlic on the spontaneous formation of N-nitrosomorpholine. Control incubations resulted in the formation of 7.4 mM NMOR after 1 h. Values are means ± SEM of 3 observations per treatment. Values occurring following addition of garlic, elephant garlic or onion extracts differ signficantly from control values (P<0.05). The pooled SEM for all measurements was ± 0. 79. **(Reproduced with permission from reference 15. Copyright 1997 Lawrence Associates, Inc.)**

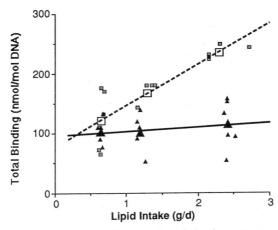

Figure 2. Effect of increasing dietary corn oil in the presence or absence of supplemental garlic on the *in vivo* total binding of DMBA to mammary cell

DNA. Rats were fed diets containing 0 (□) or 20 g garlic/kg diet (▲) for 2 weeks before DMBA treatment (25 mg/kg bw). Values represent individual (small symbol) and means (large symbol) ± SE adducts for 5 rats per treatment. **(Reproduced with permission from reference 12. Copyright 1996 American Institute of Nutrition.)**

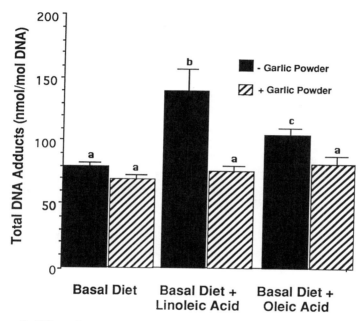

Figure 3. Effect of garlic supplementation on the occurrence of DMBA induced adducts in rats fed a 5% corn oil based diet with or without 156 mmol/kg linoleic or oleic acids. Both linoleic and oleic acid supplementation significantly increased the DMBA-DNA binding. Garlic blocked the increase in adducts caused by either fatty acid supplement.

150

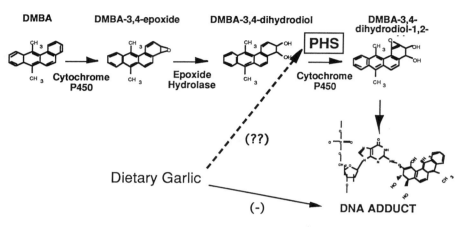

Prostaglandin H Synthase-Mediated Formation of 7,12-dimethylbenz(a)anthracene (DMBA)-DNA

DMBA   DMBA-3,4-epoxide   DMBA-3,4-dihydrodiol   DMBA-3,4-dihydrodiol-1,2-

Cytochrome P450   Epoxide Hydrolase   PHS   Cytochrome P450

(??)

Dietary Garlic

(-)   DNA ADDUCT

Figure 4. Possible involvement of prostaglandin synthase activity in the carcinogenic bioactivation of DMBA.

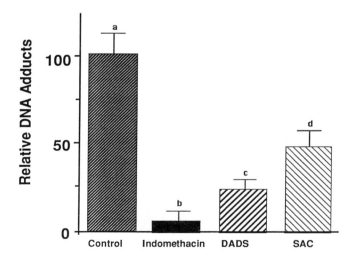

Figure 5. Influence of indomethacin, SAC and DADS on prostaglandin synthase (PHS) mediated bioactivation of DMBA. Incubations contained 400 units PHS, 1 mg/ml calf thymus DNA, 50 μM DMBA- 3,4 dihydrodiol, 100 μM arachidonic acid and 100 μM inhibitor (indomethacin, SAC or DADS) in phosphate buffer (pH 7.6). Incubations were for 10 minutes at 37° C. Indomethacin, SAC and DADS significantly (P<0.05) decreased DNA adduct formation.

Summary and Conclusions

Overall, a variety of investigations demonstrate that garlic and/or its components inhibit the cancer process at several sites. These investigations reveal that the benefits of garlic are not limited to a specific species, to a particular tissue, or to a specific carcinogen. The diversity of garlic's anti-carcinogenic effects indicates potential influence on human cancers and may explain part of the international variation in cancer incidence. Only after more extensive investigations can one predict how beneficial these compounds might be in reducing human cancer risk. Few foods are recognized to have such widespread protective effects against the cancer process as have been observed with garlic.

References

1. Wynder, E.L.; Gori, G.B. *J. Natl. Cancer Inst.* **1977**, 58, 825-832.
2. Lea, M.A. *Adv. Exp. Med. Biol.* **1996**, 401, 147-154.
3. Milner, J.A. *Nutr. Rev.* **1996**, S-82-86.
4. You, W.C.; Blot, W.J.; Chang, Y.S.; Qi, A., Henderson; B.E.; Fraumeni, J.F.; Wang, T.G. *J. Natl. Cancer Inst.* **1989**, 81, 162-164.
5. Witte, J.S.; Longnecker, M.P.; Bird, C.L.; Lee, E.R.; Frankl, H.D.; Haile, R.W. *Am. J. Epidemiol.* **1996**, 144(11), 1015-1025.
6. Steinmetz, K.A.; Kushi, L.H.; Bostick; R.M.; Folsom, A.R.; Potter, J.D. *Am. J. Epidemiol.* **1994**, 139(1), 1-15.
7. Zheng, W.; Blot, W.J.; Shu, X.O.; Gao, Y.T; Ji, B.T.; Ziegler, R.G.; Fraumeni, J.F. Jr. *Am. J. Epidemiol.* **1992**, 136(2), 178-191.
8. Mei, X.; Wang, M.L.; Xu, H.X.; Pan, X.Y.; Gao, C.Y.; Han, N.; Fu, M.Y. *Acta Nutrimenta Sinica.* **1982**, 4, 53-56.
9. Dorant, E.; van den Brandt, P.A.; Goldbohm, R.A. *Carcinogenesis.* **1996**, 17(3), 477-484.
10. Dorant, E.; van den Brandt, P.A.; Goldbohm, R.A. *Breast Cancer Res. Treatment.* **1995**, 33(2), 163-70.
11. Huang, C; Zhang, X; Qiao, Z; Guan, L.; Peng, S; Liu, J; Xie, R; Zheng, L. *Biomed. Environ. Sci.* **1992**, 5(3), 257-65.
12. Amagase, H; Schaffer, E.M.; Milner, J.A. *J. Nutr.* **1996**, 126, 817-824.
13. Farbman, K.S.; Barnett, E.D.; Bolduc, G.R.; Klein, J.O. *Pediat. Infectious Dis. J.* **1993**, 12(7), 613-614.
14. Mei, X.; Wang, M.L.; Han, N. *Acta Nutrimenta Sinica.* **1985**, 7, 173-176.
15. Dion, M.E.; Agler, M.; Milner, J.A. *Nutr. Cancer.* **1997**, 28, 1-6.
16. Shenoy, N.R.; Choughuley, A.S. *Cancer Lett.* **1992**, 65, 227-232.
17. Lin, X.Y.; Liu, J.Z.; Milner, J.A. *Carcinogenesis.* **1994**, 15, 349-352.
18. Brady, J.F.; Wang, M.H.; Hong, J.Y.; Xiao, F.; Li, Y.; Yoo, J.H.; Ning, S.M.; Lee, M.J.; Fukuto, J.M.; Gapac, J.M.; Yang, C.S. *Toxic. Appl. Pharmcol.* **1991**, 108, 342-354.
19. Peter, R.; Bocker, R.; Beaune, P.H.; Iwasaki, M.; Guengerich, F.P.; Yang, C.S. *Chem. Res. Toxicol.* **1990**, 3, 566-573.
20. Dion, M.E.; Milner, J.A. *FASEB J.* 1997, 10(3), A498.
21. Wargovich, M.J.; Woods, C., Eng, V.W.S.; Stephens, L.C.; Gray, K.N. *Cancer Res.* **1988**, 48, 6872-6875.
22. Hatono, S.; Jimenez, A.; Wargovich, M.J. *Carcinogenesis.* **1996**, 17:1041-1044.
23. Liu, J.Z.; Lin, R.I.; Milner, J.A. *Carcinogenesis.* **1992**, 13, 1847-1851.
24. Sumiyoshi, H.; Wargovich, M.J. *Cancer Res.* **1990**, 50, 5084-5087.
25. Perchellet, J.P.; Perchellet, E.M.; Belman, S. *Nutr. Cancer.* **1990**, 14, 183-193.

26. Amagase, H.; Milner, J.A. *Carcinogenesis.* **1993**, 14, 1627-1631.
27. Schaffer, E.M.; Liu, J.Z.; J.A. Milner. *Nutr. Cancer.* **1997**, 27, 162-168.
28. Ip, C.; Lisk, D.J.; Stoewsand, G.S. *Nutr. Cancer.* **1992**, 17, 279-286.
29. Schaffer, E.M.; Milner, J.A. *Cancer Lett.* **1996**, 106, 177-183.
30. (Schaffer, E. M.; Milner, J.A. *Proc. 16th Internat. Congress of Nutr.*, in press.)
31. Eling, T.E.; Curtis, J.F. *Pharmacol. Therapeutics.* **1992**, 53, 261-273.
32. Ara, G.:Teicher, B.A. *Prost. Leuk. Ess. Fatty Acids.* **1996**, 54, 3-16.
33. Schaffer, E.; Liu, J.Z.; Green, J.; Dangler, C.A.; Milner, J.A. *Cancer. Lett.* **1996**, 102, 199-204.

Chapter 15

Antioxidative and Antitumorigenic Properties of Rosemary

Chi-Tang Ho[1], Jianhong Chen, Guangyuan Lu[1], Mou-Tuan Huang[2], Yu Shao[1,3], and Chee-Kok Chin[3]

[1]Department of Food Science, [2]Laboratory for Cancer Research, and [3]Department of Plant Science, Rutgers University, New Brunswick, NJ 08903

Rosemary leaves have potent antioxidant activity. They have been widely used as a spice and an antioxidant in food processing. Rosemary inhibits skin, colon, and mammary carcinogenesis. Rosemary, and its constituents carnosic acid and carnosol, inhibit B[a]P-induced increases in cytochrome P4501A1 activity and induce increases in the detoxification enzymes glutathione S-transferase and NADPH quinone oxidoreductase activity. Carnosol also inhibits the growth and the synthesis of DNA and RNA in human leukemia HL-60 cells in a dose dependent manner.

Rosemary (*Rosmarinus officinalis* L.) and sage (*Salvia officinalis*) leaves are commonly used as spices and flavoring agents. The dried leaves of rosemary are one of the most widely used spices in processed foods because it has a desirable flavor and a high antioxidant activity (*1-3*). Crude and refined extracts of rosemary are now commercially available for application in food stabilization (*4*).

Antioxidant Activity of Phenolic Diterpene in Rosemary

Many antioxidants in rosemary are nonvolatile, and the antioxidant activity of rosemary in lard is comparable to that of BHA and BHT (*5-6*). Several diterpenes, such as carnosol (**1**), carnosic acid (**2**), rosmanol (**3**), epirosmanol (**4**), isorosmanol (**5**), rosmaridiphenol (**6**), rosmadial (**7**) and miltirone (**8**) (Figure 1) with antioxidant activities have been isolated from rosemary (*1,6-10*). Carnosol and carnosic acid account for over 90% of the antioxidative activity of rosemary (*4*). The antioxidative activities of carnosol and carnosic acid were reported as early as in 1960's (*11*). The antioxidant activity of carnosol was confirmed later by a detailed study by Wu et al.

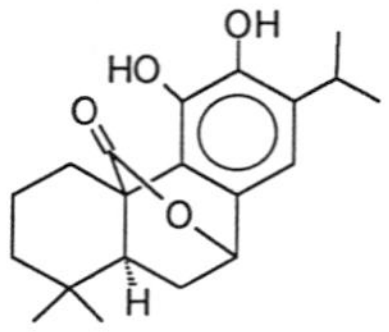

(1) Carnosol

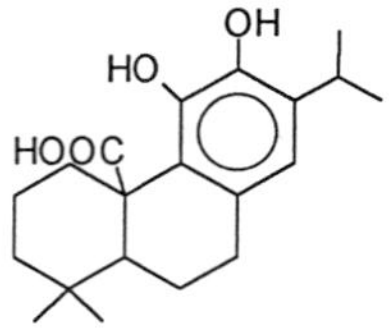

(2) Carnosic acid

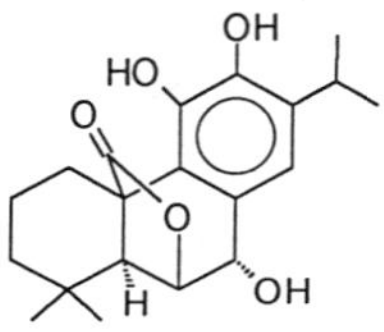

(3) Rosmanol

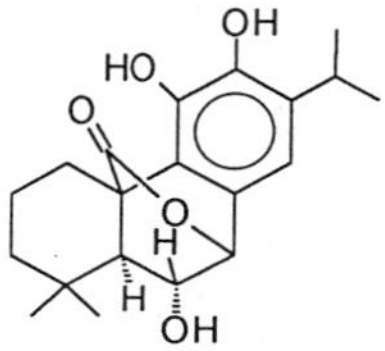

(4) Epirosmanol

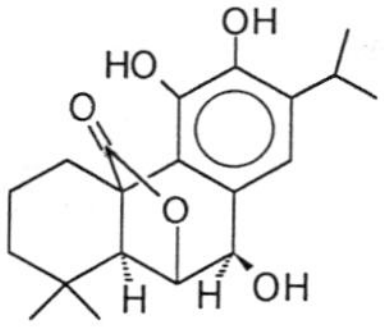

(5) Isorosmanol

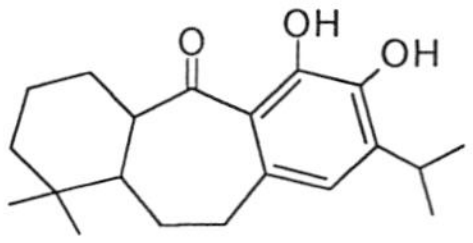

(6) Rosmanridiphenol

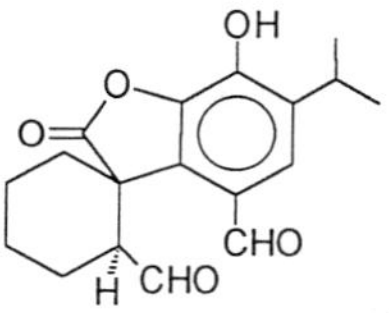

(7) Rosmadial

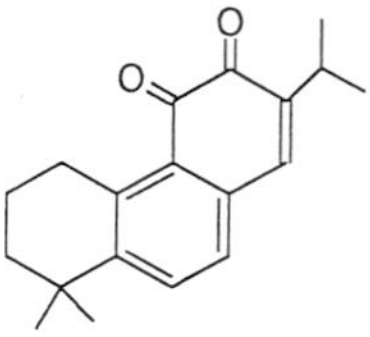

(8) Miltirone

Figure 1. Phenolic diterpenes identified in rosemary.

(*1*). The studies by Nakatani's group showed that rosmanol, epirosmanol, isorosmanol and carnosol had remarkably high activity when evaluated in lard using the Active Oxygen Method (*10,12*). Among them, rosmanol, epirosmanol, and isorosmanol were more than four times as effective as BHT, and carnosol was twice as active (*10,12*). In an aqueous ethanoic solution, the results by the ferric thiocyanate method indicated that the antioxidative activity of rosmanol and carnosol was stronger than α-tocopherol, but weaker than BHT (*10*). Rosmaridiphenol isolated by Houlihan et al. (*8*) has been shown to have antioxidative activity superior to that of BHA in lard.

Antioxidant Activity of Flavonoids in Rosemary

Generally, flavonoids are a major group of antioxidants in plants and their antioxidative activities have been widely reported (*13-14*). However, only a few of the flavonoids found in rosemary possess antioxidative activity. Luteolin is one of those flavonoids, which has weaker antioxidative activity than quercetin (*14*). Another flavonoid, 5-hydroxy-7,4'-dimethoxyflavone isolated by Inatani et al. (*12*) has shown antioxidative activity. However, its antioxidative activity is weaker than that of rosmanol. Compared to synthetic antioxidants, such as BHA and BHT, the antioxidative activity of these flavonoids is relatively weak. The lack of or low antioxidative activity of this group of compounds may be attributed to the fact that there is no hydroxyl group on the C_3 position of the molecule (*14*). Recently, Okamura et al (*15*) isolated four flavonoid glycosides, 5-hydroxy-7,4'-dimethoxy flavone, luteolin 3'-*O*-β-D-glucuronide (**10**), luteolin 3'-*O*-(4"-O-acetyl)-β-D-glucuronide (**11**), Luteolin 3'-*O*-(3"-O-acetyl)-β-D-glucuronide (**12**), and hesperitin. The chemical structures of compounds, **10**, **11** and **12** are shown in Figure 2. The antioxidative activities of these flavonoids were evaluated by the ferric thiocyanate method. All four compounds were active and the most active compound was hesperitin.

Other Compounds in Rosemary

Rosemary is known to contain a significant quantity of rosmarinic acid which is a dimer of caffeic acid. Ursolic acid, a triterpene acid, is also a well-known component of rosemary. Ursolic acid is not an active antioxidant (*1*), but its biological activity has been explored (*16*). Recently, we have isolated a new lingan glycoside, rosmarinoside A (**14**) from the more polar fraction of rosemary leaf extract (Shao, Y.; Ho, C.-T.; Chin, C. K., unpublished results). The antioxidant and biological activity of the lignan glycoside is under investigation. The chemical structures of rosmarinic acid, ursolic acid and rosmarinoside A are shown in Figure 2.

Free Radical Quenching Activity of Rosemary

The mechanism for the antioxidant activity of rosemary constituents has not been extensively studied. The free radical scavenging activities were tested by Zhao et al.

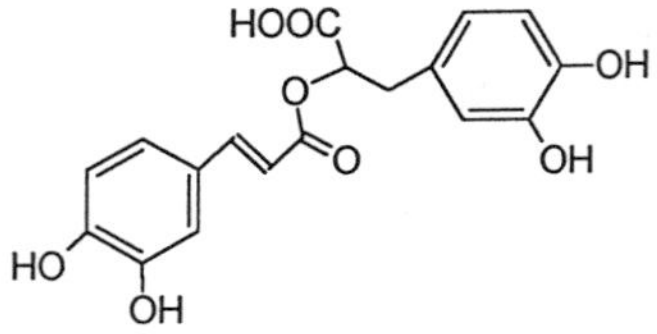

(9) Rosmarinic Acid

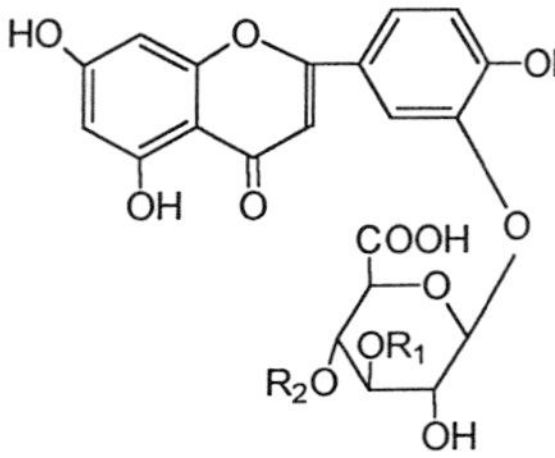

(10) Luteolin 3'-O-β-D-glucuronide R_1=H, R_2=H
(11) Luteolin 3'-O-(4''-O-acetyl)-β-D-glucuronide R_1=H, R_2=Ac
(12) Luteolin 3'-O-(3''-O-acetyl)-β-D-glucuronide R_1=Ac, R_2=H

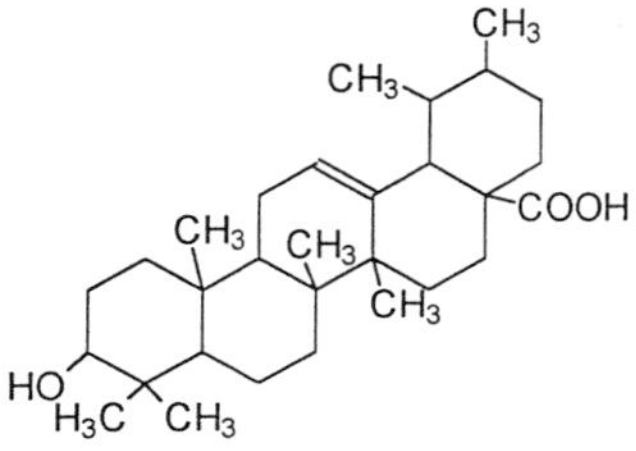

(13) Ursolic acid 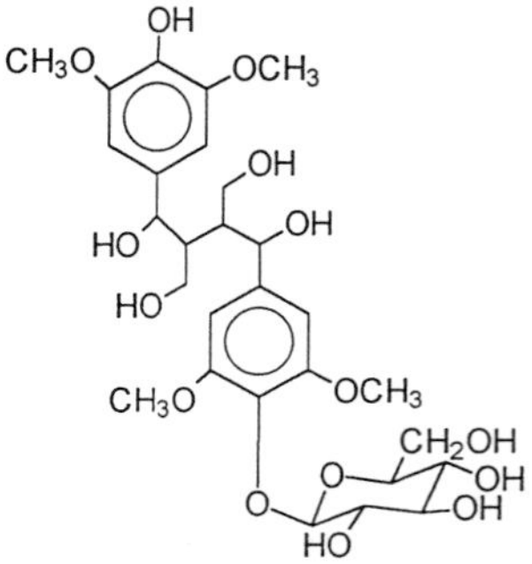   (14) Rosmarinoside A

Figure 2. Rosmarinic acid, flavonoid glycosides, ursolic acid and a new lignan identified in rosemary.

(*17*) in several reaction systems. They found that the scavenging activity of rosemary antioxidants was dependent upon the species of free radicals. Rosemary antioxidants scavenged the active oxygen radical more effectively than α-tocopherol, but less effectively than ascorbic acid in a stimulated polymorphonuclear leukocytes system. Similarly, the scavenging effect of rosemary antioxidants on superoxide anion was stronger than that of α-tocopherol, but weaker than ascorbic acid in an irradiated riboflavin reaction system. In a Fenton reaction system, no scavenging effect of rosemary antioxidants on hydroxyl radicals was found. It should be noted that the composition of rosemary antioxidants used by Zhao et al. (*17*) was not well defined. The free radical scavenging activity of the individual constituents of rosemary has recently been tested by Aruoma et al. (*18*). They found that carnosol and carnosic acid can effectively scavenge peroxy radicals generated by radiolysis of a mixture of propan-2-ol and carbon tetrachloride, and also hydroxyl radicals in a deoxyribose assay. In this same study (*18*) the iron chelating effect of carnosol and carnosic acid was also found.

Antitumorigenic Activity of Rosemary Extract

Like many other antioxidants, rosemary or some of its components possess not only antioxidative activities, but also antitumorigenic activities. Huang et al. (*19*) tested the inhibitory effects of rosemary extract, carnosol, and ursolic acid on the tumor formation in mouse skin. They found that topical application of rosemary inhibits benzo[a]pyrene (B[a]P)- and 7,12-dimethylbenz[a]anthracene (DMBA)-induced initiation of tumors and 12-*O*-tetradecanoylphorbol-13-acetate (TPA)-induced tumor promotion in DMBA-initiated mice. The mechanism for the rosemary to inhibit the initiation of tumor by B[a]P is that a topical application of rosemary inhibits covalent binding of B[a]P to epidermal DNA. With regard to the inhibitory effect of rosemary on the tumor formation initiated by DMBA, Singletary and Nelshoppen (*20*) reported the same mechanism. They fed the rosemary extract to the rats treated orally with DMBA and found that it inhibited the covalent binding of DMBA to total DNA and to deoxyguanosine, but not to deoxyadenosine, in the mammary gland. Other effects of rosemary included strong inhibition of TPA-induced increases in ornithine decarboxylase activity, inflammation, hyperplasia, and tumor promotion in mouse skin (*19*). Rosemary in the diet was found to inhibit B[a]P-induced forestomach and lung tumorigeneis, azoxymethane-induced colon tumorigenesis, and DMBA-induced mammary gland tumorigenesis in mice (*21*).

Antitumorigenic Activity of Rosemary Constituents

Carnosol and ursolic acid are the two major components of rosemary. They were also tested and found to be strong inhibitors of TPA-induced inflammation, ornithine decarboxylase activity, and tumor promotion in mouse skin (*19*). Carnosol was also found to inhibit mammalian 5-lipoxygenase activity and arachidonic acid-induced inflammation (*19*). It was suggested that carnosol acted like other nonsteroidal phenolic anti-inflammatory agents such as curcumin which inhibited the metabolism of arachidonic acid. Ursolic acid, a non-antioxidant in rosemary was also found to be a potent inhibitor of TPA-induced inflammation and tumor promotion in mouse skin (19,22). Other inhibitory effects of ursolic acid were found on TPA-induced Epstein-Barr virus activation in Raji cells (22-23), and 12-O-hexadecanoyl-16-hydroxyphorbol-13-acetate-induced edema on mouse ears (24). Compared to rosemary extract, ursolic acid or carnosol alone has less inhibitory effect on TPA-induced tumor promotion (21). It is uncertain whether rosemary extract possesses a synergistic effect by these two compounds or there are some other more potent compounds in rosemary extract.

Effect of Rosemary on Phase I and Phase II Enzymes

An *In vitro* human cell model was developed by Offord et al. (4,25) to study the mechanisms by which rosemary and its components exert their anticarcinogenic effect. They used human bronchial epithelial (BEAS-2B) cells to study the inhibitory effect of carnosol, carnosic acid, and rosemary extract on the formation of DNA adducts by activated metabolites of B[a]P. They reported that at least two mechanisms were involved in the inhibition of the DNA adduct formation; the inhibition of the phase I enzyme cytochrome P4501A1 and the induction of the phase II enzyme glutathione-S-transferase. Carnosol and carnosic acid also induced the phase II enzyme quinone reductase which catalyses the detoxification of reactive quinone intermediates (25).

Inhibitory Activity of Carnosol Against Human Leukemia HL-60 Cells in Culture

We have studied the effects of carnosol and carnosic acid against macromolecular biosynthesis and cellular growth of human leukemia HL-60 cells. These experiments were performed as previously described (26).

Carnosol effectively arrested the proliferation of HL-60 cells in a dose-dependent manner (Figure 3). It exhibited cytocidal activity at a concentration greater than 10 μM, and inhibited HL-60 cell growth compared with the control when a concentration of carnosol was less than 10 μM. In 48 hours, HL-60 cell growth was arrested by 13, 33, 39, 58 and 97% at 2.0 μM, 5.0 μM, 7.5 μM, 10 μM and 50 μM carnosol respectively.

As shown in Figure 4, carnosol inhibited DNA, RNA and protein synthesis in HL-60 cells; the IC_{50} were determined to be 7.0 μM, 5.0 μM and 17 μM respectively. Their syntheses were completely inhibited at around 50 μM carnosol.

Carnosic acid has a weaker inhibitory effect than carnosol. The addition of 20 μM of carnosic acid induced only 20% inhibition on DNA synthesis in HL-60 cells.

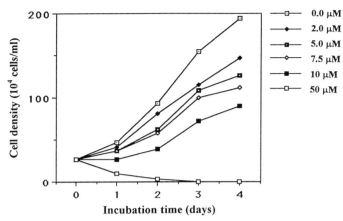

Figure 3. Inhibitory effect of carnosol on HL-60 cell growth. Fifteen μl DMSO solutions with various concentrations of carnosol or 15 μl DMSO (for control) were added in a series of culture dishes containing HL-60 cells in 15 ml RPMI-1640 medium with 10% FCS, 100 μu/ml penicillin and 100 μg/ml streptomycin. Cell cultures were incubated at 37°C in 5% CO_2 humidified atmosphere for four days. The numbers of HL-60 cells were counted every 24 hours.

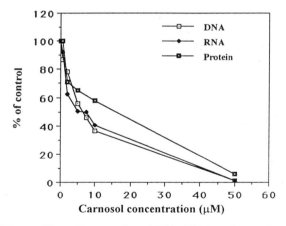

Figure 4. Inhibitory effect of carnosol on DNA, RNA and protein synthesis of HL-60 cells. Two μl DMSO solution of various concentrations of carnosol or 2 μl DMSO (for control) were added in a series of culture tubes containing HL-60 cells in one ml RPMI-1640 medium without FCS, penicillin and streptomycin. Cell concentration was 4×10^5 cells/ml. ^{3}H labeled thymidine (for monitoring DNA synthesis), or uridine (for monitoring RNA synthesis) or leucine (for monitoring protein synthesis) were added, the amount were 3 μl/ml, 5 μl/ml and 8 μl/ml respectively. Tubes were incubated at 37°C for 100 minutes. The rates of DNA, RNA and protein synthesis were determined as described under Methods. The ^{3}H radioactivity ratio of samples to control was calculated as percentage of control.

160

Literature Cited

1. Wu, J. W.; Lee, M-H.; Ho, C-T.; Chang, S. S. *J. Amer. Oil Chem. Soc.* **1982**, *59*, 339-345.
2. Ho, C.-T.; Ferraro, T.; Chen, Q.; Rosen, R. T.; Huang, M. T. in *Food Phytochemicals for Cancer Prevention II. Teas, Spices, and Herb;* Ho, C.-T.; Osawa, T.; Huang, M. T.; Rosen, R. T. , Eds.; ACS Symp. Ser. No. 547; American Chemical Society: Washington, D.C., 1994; pp. 2-19.
3. Nakatani, N. in *Natural Antioxidants: Chemistry, Health Effects, and Application*; Shahidi, F. Ed.; AOCS Press: Champaign, IL, 1997, pp. 64-75.
4. Offord, E. A.; Guillot, F.; Aeschbach, R.; Loliger, J.; Pfeifer, A. M. A. in *Natural Antioxidants: Chemistry, Health Effects, and Application*; Shahidi, F. Ed.; AOCS Press: Champaign, IL, 1997, pp. 88-96.
5. Chang, S. S.; Ostric-Matijasvic, B.; Hsieh, O.A.L.; Huang, C-L. *J. Food Sci.*, **1977**, *42*, 1102-6.
6. Nakatani, N.; Inatani, R. *Agric. Biol. Chem.* **1981**, *45*, 2385-2386.
7. Inatani, R.; Nakatani, N.; Fuwa, H.; Seto, H. *Agric. Biol. Chem.* **1982**, *46*, 1661-1666.
8. Houlihan, C.M.; Ho, C.-T.; Chang, S. S. *J. Amer. Oil Chem. Soc.* **1984**, *61*, 1036-1039.
9. Houlihan, C.M.; Ho, C.-T.; Chang, S. S. *J. Amer. Oil Chem. Soc.* **1985**, *62*, 96-8.
10. Nakatani, N.; Inatani, R. *Agric. Biol. Chem.* **1984**, *48*, 2081-2085.
11. Brieskorn, C. H.; Domling, H. J. Z. *Lebensm. Unters. Forsch.* **1969**, *141*, 10-16.
12. Inatani, R.; Nakatani, N.; Fuwa, H. *Agric. Biol. Chem.* **1983**, *47*, 521-528.
13. Kandaswami, C.; Middleton, E. in *Natural Antioxidants: Chemistry, Health Effects, and Application*; Shahidi, F. Ed.; AOCS Press: Champaign, IL, 1997, pp. 174-203.
14. Ho, C.-T. in *Food Factors for Cancer Prevention*; Ohigashi, H.; Osawa, T.; Terao, S.; Watanabe, S.; Yoshikswa, T., Eds.; Springer-Verlag Tokyo: Tokyo, 1997, pp. 593-597.
15. Okamura, N.; Haraguchi, H.; Hashimoto, K.; Yagi, A. *Phytochem.* **1994**, *37*, 1463-1466.
16. Huang, M. T.; Ho, C.-T.; Ferraro, T.; Wang, Z. Y.; Stauber, K.; Georigiadis, C.; Laskin, J. D.; Conney, A. H. *Proc. Am. Assoc. Cancer. Res.* **1992**, *33*, 165.
17. Zhao, B.; Li, X; He, R.; Cheng S.; Xin, W. *Cell Biophys.* **1989**, *14*, 175-185.
18. Aruoma, O.I.; Halliwell, B.; Aeschbach, R.; Loliger, J. *Xenobiotics* **1992**, *22*, 257-268.
19. Huang, M.-T.; Ho, C-T.; Wang,, Z. Y.; Ferraro,, T.; Lou,Y-R; Stauber, K.; Ma, W.; Georgiadis, C.; Laskin, J.D.; Conney, A. H. *Cancer Res.* **1994**, *54*, 701-708.
20. Singletary, K. W.; Nelshoppen, J. M. *Cancer Lett.* **1991**, *60*, 169-175.

21. Huang, M. T.; Ho, C.-T. in *Food Factors for Cancer Prevention*; Ohigashi, H.; Osawa, T.; Terao, S.; Watanabe, S.; Yoshikswa, T., Eds.; Springer-Verlag Tokyo: Tokyo, 1997, pp. 253-256.

22. Tokuda, H.; Ohigashi,H.; Koshimizu, K.; Ito, Y. *Cancer Lett.* **1986**, *33*, 279-285.

23. Ohigashi, H.; Takamura, H.; Koichi, K; Tokuda, H.; Ito, Y. *Cancer Lett.* **1986**, *30*, 143-151.

24. Hirota, M.; Mori T.; Yoshida, M.; Iriye, R. *Agric. Biol. Chem.* **1990**, *54*, 1073-1075.

25. Offord, E. A.; Mace, K.; Ruffieux, C.; Malnoe, A.; Pfeifer, A. M. A. *Carcinogenesis* **1995**, *16*, 2057-2062.

26. Chen, J. H.; Shao, Y.; Huang, M. T.; Chin, C. K.; Ho, C.-T. *Cancer Lett.* **1996**, *108*, 211-214.

Chapter 16

Nontoxic and Nonorgan Specific Cancer Preventive Effect of *Panax ginseng* C. A. Meyer

Taik-Koo Yun[1,3], Soo-yong Choi[2], and Yun-Sil Lee[1]

[1]Laboratory of Experimental Pathology and [2]Laboratory of Clinical Research, Korea Cancer Center Hospital, Seoul 139-240, Korea

Panax ginseng C. A. Meyer has been recognized as non-toxic mysterious tonic in the Orient. The prolonged administration of red ginseng extract inhibited the incidence and also proliferation of pulmonary tumors induced by various chemical carcinogens. Statistically significant anticarcinogenic effects were observed in powders and extract of 6 year-dried fresh ginseng, 5 and 6 year-white ginseng and 4, 5 and 6 year-red ginseng by newly established 9 week medium-term anticarcinogenicity test using benzo[a]pyrene (BP). In case-control studies, odds ratios (OR) of the cancer of lip, oral cavity and pharynx, larynx, lung, esophagus, stomach, liver, pancreas, ovary, and colorectum were significantly reduced. On the type of ginseng, the ORs for cancer were reduced in fresh ginseng extract intakers, white ginseng extract intakers, white ginseng powder intakers, and red ginseng intakers. In cohort study with 5 years follow-up conducted in ginseng cultivation area, ginseng intakers had a decreased relative risk (RR) compared with non-intakers. The RRs of ginseng intakers were decreased in gastric cancer and lung cancer. These findings strongly suggest that *Panax ginseng* C. A. Meyer in non-toxic and non-organ specific cancer preventive effects against various cancers. Ginseng should be recognized as a functional food for cancer prevention, and that further studies for the identification of its active components, mechanism of action and clinical interventions should undertaken with worldwide collaboration.

Rapid advances in scientific and medical knowledge, coupled with the rise in health consciousness of the population and increased emphasis on health promotion, have focused worldwide attention on food as on important component in the prevention

---

[3]Current address: 215-4 Gongneung Dong, Nowon Ku, Seoul 139-240, Korea.

of diseases. On the other hand, fifty years already passed since alkylating agent, nitrogen mustard was developed as first cancer chemotherapeutics (1), but many cancers still remain difficult to cure (2). With the discovery and clinical application of cancer chemotherapeutics, five year survival rate, which was less than one in five in the 1930s increased to one in four in the 1950s and one in three in the 1960s. By the 1970s, the goal of five year cancer survival rate was to achieve one out of two patients, but it was not successful. At present, it is only two out of five patients corresponding to 40% of "observed" survival rate (3).

It has been believed that cancer should be conquered by prevention since 1978, and that it is not desirable the use synthetic agents for chemoprevention due to their toxicity problems. Therefore, I have been trying to discover non-toxic cancer preventives in natural products. The necessity of developing new preventives for chemoprevention or immunoprevention from natural products, which we have been taking for a long time, was also understood.

We hypothesized that the life-prolongation effect of ginseng described by Shennong (4) in Liang Dynasty China may due to the preventive activity of ginseng against the development of cancers. Our study has focused on whether ginseng has anticarcinogenicity against various chemical carcinogens, such as urethane, 9, 10-dimethyl-1,2-benzanthracene (DMBA), N-2-fluorenylacetamide (FAA), N-methyl-N'-nitroso-N-nitroguanine (MNNG), aflatoxin B_1 and tobacco smoke condensates for long-term period maximum 67 weeks (5,6).

Thereafter, we established new 9 week medium-term anticarcinogenicity model (termed Yun's anticarcinogenicity test) using one of the environmental carcinogens, BP at our laboratory to conform the anticarcinogenicity of red ginseng (7-10) and compare the anticarcinogenicity of various types and ages of ginseng (11, 12).

In 1987 we began to conduct an epidemiological study to confirm whether red ginseng extracts have as much anticarcinogenic effect on human beings as on mice. In relation to this work we made three studies; two case-control studies on cancer patients (13, 14) and a cohort study on a population of a ginseng cultivation area (15). Here, experimental and epidemiological evidence of the preventive effect of ginseng is reviewed.

Nomenclature and types of ginseng

The species of ginseng are *Panax ginseng* C. A. Meyer (Korean ginseng), which is cultivated in Korea, Japan, China and Russia; *Panax quinquefolius* L.(American ginseng), which is raised in the eastern United States and Canada; *Panax japonicus* C. A. Meyer (Japanese ginseng), which is also called Bamboo ginseng; and *Panax notoginseng* (Burk) F. H. Chen (Sanchi-ginseng), a native of southwest China (Yunnan and Kwangsi Provinces). In Korea, *Panax ginseng* C. A. Meyer is collected after 2 to 6 years of cultivation, and it is classified into three types depending on how it is processed: (a) fresh ginseng (less than 4 years old and can be consumed in the fresh state); (b) white ginseng (4-6 years old and then dried after peeling); and (c) red ginseng, which is harvested when 6 years old and then steamed and dried. Each type of ginseng was categorized further into several forms of ginseng products, fresh sliced, juice, extract (tincture or boiled extract), powder, tea, tablet, capsule, and other forms (Fig. 1).

Long-term anticarcinogenicity experiment

This investigation was carried out to evaluate the effects of ginseng in inhibition or prevention of carcinogenesis induced by various chemical carcinogens. Red ginseng extract (1mg/ml of drinking water) was administered orally to the weaned mice. Chemical carcinogens, 9,10-demethyl-1, 2-benzanthracene (DMBA, 30μg), urethane (1mg), N-2 fuorenylacetamide (FAA, 100μg x 5), aflatoxin B_1 (8μg),and Hansando tobacco smoke condensates (320μg) were injected in the subscapular region of ICR mice within 24 hours after birth. Controls comprised three groups of ICR newborn mice: normal (100), ginseng (200), and vehicle (316). The ten experimental groups of ICR newborn–mice comprised DMBA (101), DMBA combined with ginseng (103), urethane (94), urethane combined with ginseng (92), FAA (90), FAA combined with ginseng (88), aflatoxin B_1 (50), aflatoxin B_1 combined with ginseng (47). In the MNNG group, MNNG (3mg) was injected subcutaneously into the backs of Wistar rats once a week for 10 weeks (5). The mice and rats autopsied immediately following sacrifice. All major organs were examined grossly and weighed. Histopathological examinations were also made. In the–group sacrificed at 48 weeks after the treatment with DMBA (DMBA combined with ginseng), the incidence of diffuse infiltration of pulmonary adenoma decreased by 61% (p<0.01), and the average lung weight of male mice decreased by 21% (p<0.05). In the group sacrificed at 28 weeks after the treatment (urethane combined with ginseng), there was 22% decrease (p<0.05) in the incidence of lung adenoma. In the group sacrificed at 56 weeks after birth (aflatoxin B1 combined with ginseng), there were decrease in the incidence of lung adenoma (29%) (p<0.05) and hepatoma (75%) (p<0.05) (Table 1). In the groups sacrificed 68 weeks after the treatment with FAA and 48 weeks of tobacco smoke condensate, statistically no significant decrease was observed. In the group sacrificed 27 weeks after treatment with MNNG (MNNG combined with ginseng), ginseng extract had no effect on the incidence of sarcoma induced by MNNG. These findings indicate that prolonged administration of red ginseng extract inhibited the incidence and also the proliferation of tumors induced by DMBA, urethane, and aflatoxin B_1 (5).

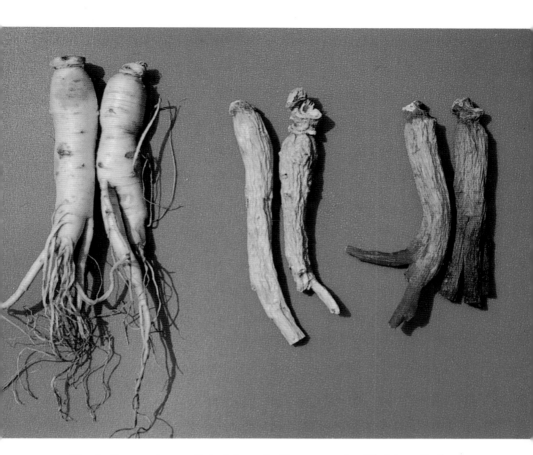

Fig 1. *Panax ginseng* C.A. Meyer in Korea are classified into fresh ginseng (left), less than 4 years old, and can be consumed in the fresh state; white ginseng (center), 4-6 years old and then dried after peeling; and red ginseng (right), harvested when 6 years old and then steamed and dried

Establishment of medium-term anticarcinogenicity test (Yun's test)

As mentioned previously, we performed a long-term experiment for 67 weeks. But we realized that it was necessary to develop a medium-term model for further experiments covering 1) fluctuation of experimental conditions due to the long-term feeding, 2) masking or paving the inhibitory activities of early stages, and 3) usage of synthetic environmental chemical carcinogens such as BP. Therefore, we decided to establish a 9-12 weeks medium-term anticarcinogenicity test model (7). Newborn mice less than 24 hours old of A/J, C57BL/6J, C57BR/cdJ and N:GP(S) strains were injected subcutaneously with 0.5 mg or 1 mg of BP. All mice were sacrificed at the 9th week after birth. Lungs were excised and fixed in Tellyesniczky's solution (100ml of 70% ethanol, 3ml of formalin, 5ml of glacial acetic acid). Then the adenoma were counted by the naked eye. After counting the lungs were embedded in paraffin, then cut and stained with hematoxylin-eosin. To obtain an index of tumor incidence, the percentage of tumor bearing mice per total number of mice in each group was calculated. Tumor multiplicity was defined as the average number of tumors per mouse obtained by dividing the total number of tumors by the total number of mice per group including nontumor-bearing animals. Statistical comparison were made using the Chi-square test for tumor incidence and Student(s t-test of multiplicity. A null hypothesis was rejected whenever a P value of 0.05 or less was found. Lung adenoma incidence was 46.8% and 54.4% in N:GP(S) mice at a concentration of 0.5 mg and 1 mg in BP treated groups, respectively. That of A/J mice was 86.7% and 88.3%, that of C57BL/6J mice was 1.3% and 0%, and that of C57BR/cdJ was 0% (Table 2). The dose response effect of BP in A/J and N:GP(S) mice was also examined. A single injection of 40 μg of BP, which was the lowest dose in this experiment, showed an incidence of lung tumors of 71.0% in A/J mice, which might be too high in incidence to evaluate the anticarcinogenicity of unknown compounds. However, the dose showing a 50% tumor incidence was found in N:GP(S) mice to be 0.5 mg of BP (49.4%) (7, 9,10).

Evaluation of the new anticarcinogenicity test model established

A new medium-term *in vivo* model was tried using pulmonary adenoma induced by BP in N:GP(S) newborn mice. To verify the utility of this model, ascorbic acid, carrot, beta carotene, soybean lecithin, spinach, *Sesamum indicum*, *Ganoderma lucidum*, caffeine, red ginseng extract (6 years old), fresh ginseng (4 years old), and 13-cis retinoic acid, some of which are known to have anticarcinogenic activity in various animal models, were also tried in the model. Ascorbic acid, soybean lecithin, *Ganoderma lucidum*, caffeine, and red ginseng extract showed inhibition of lung tumor incidence, while fresh ginseng, carrot, beta carotene, spinach, and 13-cis

Table 1. Odds ratio of cancer and 95% confidence interval (CI) by ginseng intake.

Ginseng Intake	Cases	Controls	Odds Ratio (95% CI)
No intake of ginseng	343	231	1.00
Intake of ginseng	562	674	0.56 (0.46-0.69)
Fresh ginseng			
Fresh slice	103	94	0.74(0.53-1.04)
Juise	39	34	0.77(0.46-1.30)
Extract*	13	64	0.14(0.07-0.26)
Extract and powder*	13	26	0.34(0.16-0.70)
Extract and fresh slice*	6	22	0.18(0.07-0.49)
White ginseng			
Extract*	247	261	0.64(0.50-0.82)
Extract and fresh slice*	54	61	0.59(0.39-0.91)
Powder*	28	43	0.44(0.26-0.75)
Powder and fresh slice	10	14	0.48(0.19-1.17)
Tea	37	27	0.93(0.53-1.61)
Red ginseng			
Extract	2	3	0.45(0.05-3.32)
Others	10	25	0.27(0.13-0.53)

*P<0.01.

International J. Epidemiology (ref.13)

Table 2. Distribution of ginseng intake frequency for cases and controls by sex: odds ratio of cancer in ginseng intake frequency and 95% confidence intervals (CI)

	Males			Females		Frequency of ginseng intake
	Case	Controls	Odds ratio(95% CI)	Cases	Controls	Odds ratio(95%CI)
None	117	56	1.00	226	175	1.001-3
times/year	132	108	0.58(0.38-0.90)	111	106	0.81(0.57-
1.15)4-11 times/year	104	115	0.43(0.28-0.67)	75	103	0.56(0.39-
0.82)Once/month or more	83	157	0.25(0.16-0.39)	57	85	0.52(0.35-
0.78)Total	436	436		469	469	

Linear trend test (1 d.f.) 45.59 (P<0.00001)   3.98 (P<0.05)
(2 homogeneity test (3 d.f.) 47.28 (P<0.00001)   16.53 (P<0.001)
International J. Epidemiology (ref.13)

retinoic acid did not (8, 10). This result suggested that this model using lung tumor induced by 0.5mg of BP was useful for the screening of cancer preventive agents.

Anticarcinogenicity of types and ages of *Panax ginseng* C. A. Meyer

We have already confirmed that 6-years-old red ginseng extract had an anticarcinogenic effect using the 9 week medium term anticarcinogenicity test. We further investigated whether fresh ginseng or white ginseng has similar anticarcinogenic effects and also if their anticarcinogenic effects are related to types and ages of ginseng using the medium-term test. In this study, fresh ginseng of 1.5, 3, 4, 5, and 6 years of age was used. Fresh ginseng was dried at room temperature, finely powdered, and extracted in a water bath for 8 hrs, 3 times (yield of extract: 45%). White ginseng was processed in the same way as fresh ginseng after removal of its cortex and fine root (yield of extract: 47%). For red ginseng, fresh ginseng was steamed, dried, and processed in the same way as fresh ginseng (yield of extract: 51%). Dried fresh ginseng and red ginseng powders or extracts of 1.5, 3, 4, 5, and 6 years of age, and white ginseng powders or extracts at 3, 4, 5, and 6 years of age were used. Newborn N:GP(S) mice were given a single subcutaneous injection of 0.5 mg of BP. Various types and ages of ginseng powders at 5 mg/ml were orally administered. All the mice were sacrificed at the 9th week. The following results were obtained: 1) In the dried fresh ginseng powder treated groups, the incidence of lung adenoma induced by BP was 41.3% and its incidence was reduced to 31.2%, 30.3%, 30.7% and 27.8% (p<0.05) after co-treatment with 1.5-, 3-, 4-, 5-, and 6-years-old dried fresh ginseng powders, respectively. A statistically significant effect was observed only in 6-year-old dried fresh ginseng powder. 2) In the white ginseng powder treated groups, the incidence of lung adenoma induced by BP was 45.0% and its incidence decreased to 41.3%, 38.0%, 31.6% (p<0.05), and 25.3% (p<0.05) after co-treatment with 3-, 4-, 5-, and 6-years-old white ginseng powders, respectively. In 5- and 6-years-old white ginseng powders, anticarcinogenic effects were observed. 3) In the red ginseng powders treated groups, the incidence of lung adenoma induced by BP was 48.6% and its incidence diminished to 37.9%, 41.7%, 31.7% (p<0.05), 28.3% (p<0.05), and 25.5% (p<0.01) after co-treatment with 1.5-, 3-, 4-, 5-, and 6-years-old red ginseng powders, respectively. In 4-, 5-, and 6-years-old red ginseng powders, anticarcinogenic effect was prominent (11). The other hand, each ginseng powder was extracted and their anticarcinogenicities were investigated. Various types and ages of ginseng extracts at 2.5 mg/ml were orally administered. All the mice mere sacrificed at the 9th week. The following results were obtained. In the dried fresh ginseng extract treated group, the incidence of lung adenoma induced by BP was 63.9% and its incidence was reduced to 48.3%, 52.5%, 51.8%, 47.5%, and 44.1% (p<0.05) after co-treatment with 1.5-, 3-, 4-, 5-, and 6-years-old dried fresh ginseng, respectively. A statistically significant effect was observed only in 6-years-old dried fresh ginseng extract. The incidence of lung adenoma induced by BP on the white ginseng extract treated group was 41.3% and showed 31.0% 46.0%, 44.0%, and 26.5% (p<0.05) after co-treatment with 3-, 4-, 5-, and 6-years-old white ginseng extracts, respectively. A statistically significant effect was observed in 6-years-old white ginseng extract. In the red ginseng extract group, the incidence of lung adenoma induced by BP was 47.5% and its incidence diminished to 40.7%, 35.0%, 30.1% (p<0.05), 30.0% (p<0.05), and 26.3% (p<0.05) after co-treatment with 1.5-, 3-, 4-, 5-, and 6-years-old red ginseng extract, respectively. In 4-, 5-, and 6-years-old red ginseng extracts, the anticarcinogenic effects were statistically significant. From the above results, we concluded that a statistically significant anticarcinogenic effect was observed in extracts of 6-years-old dried fresh ginseng, 6-years-old white ginseng, and 4-, 5-, and 6-years-old red ginseng and it is suggested that the anticarcinogenicity of ginseng varies according to the type and age (11, 12).

It has recently been shown that the biomass of ginseng tincture protected Ehlrich ascites tumor cells against the mutagenic action of urea nitrosomethyl *in vitro* and *in vivo* (16). Sanchi ginseng inhibited the early antigen activation of Epstein-Barr virus in Raji cells induced by TPA and n-butyric acid, pulmonary tumorigenesis induced in mice by 4-nitroquinoline-N-oxide and glycerol (17), and ginseng inhibited the liver cancer induced by diethylnitrosamine in rats (18). It has also been reported more recently that tissue culture biomass tincture obtained from culture cells of *Panax ginseng* C. A. Meyer had a marked inhibitory effect on mammary adenocarcinoma induced by N-methyl-N-nitrosourea administration in rats (19).

Case-control studies on 905 pairs

The effect of ginseng consumption on the risk of cancer was investigated by interviewing 905 pairs of cases and controls matched by age, sex, and date of admission to the Korea Cancer Center Hospital, Seoul. Of the 905 cases 562 (62%) had a history of ginseng intake compared to 674 of the controls (75%) a statistically significant (p<0.01). The odds ratio (OR) of cancer in relation to ginseng intake was 0.56 [95 confidence interval (CI), 0.45-0.69]. Ginseng extract and powder were shown to be more effective than fresh sliced ginseng, the juice, or tea in reducing the OR (Table 1). ORs for decreasing levels of ginseng intake were 1.00, 0.58, 0.43, and 0.25 for males and 1.00, 0.81, 0.56, and 0.52 for females (Table 2). A trend test showed a significant decrease in proportion of cancer cases with increasing frequency of intake for males (p<10-5) as well as for females (p<0.05). The reliability of recall for ginseng use was assessed by interviewing one tens of randomly selected subjects twice using the same questionnaire. The overall agreement in reported ginseng use between the two interviews was 0.85, and Kappa value was 0.71 (p<0.01).

These results strongly support the hypothesis of preventive effects of ginseng on cancer suggested by previous animal experiments (13).

Case-control study on 1987 pairs

In order to explore further (a) the types of ginseng products that have the most prominent cancer preventive effect, (b) the reproducibility of the dose-response relationship, (c) the duration of ginseng consumption that has a significant preventive effect, (d) the types of cancer which can be prevented by ginseng, and (e) the effect of ginseng on cancers associated with smoking, we extended the number of subjects for a case-control study on 1987 pairs. In this study, ginseng intakers had a decreased risk (OR; 0.50; 95% CI=0.44-0.58) for cancer compared with non-intakers similar with previous 0.56. On the type of ginseng the OR for cancer were 0.37 (95% CI=0.29-0.46) for fresh ginseng extract intakers, 0.57 (95% CI=0.48-0.68) for white ginseng extract intakers, 0.30 (95% CI=0.22-0.41) for white ginseng extract intakers, 0.30 (95% CI=0.22-0.41) for white ginseng powder intakers, and 0.20 (95% CI=0.88-0.50) for red ginseng intakers. Intakers of fresh ginseng slice, fresh ginseng juice, and white ginseng tea, however, showed no decreasing risk (Table 3). There was a decrease in risk with the rising frequency and duration of ginseng intake, showing a dose-response relationship (Table 4). On the site of cancer, the odds ratio were 0.47 (95% CI=0.29-0.76) for cancer of the lip, oral cavity, and pharynx; 0.20 (95% CI=0.09-0.38) for esophageal cancer; 0.36 (95% CI=0.25-0.52) for stomach cancer; 0.42 (95% CI=0.24-0.74) for colorectal cancer; 0.48 (95% CI=0.33-0.70) for liver cancer; 0.22 (95% CI=0.05-0.95) for pancreatic cancer; 0.18 (95% CI=0.06-0.54) for laryngeal cancer; 0.55 (95% CI=0.38-0.79) for lung cancer; 0.15 (95% CI=0.04-0.60) for ovarian cancer; and 0.48 (95% CI=0.27-0.85) for other cancers (Table 5). In cancers of female breast, uterine cervix, urinary bladder, and thyroid gland, however, there was no association with ginseng intake. In cancer of the lung, lip, oral cavity and pharynx, and liver, smokers with ginseng intake showed decreased OR compared with smokers without ginseng intake. These findings support the view that ginseng intakers had a decreased risk for most cancer compared with nonintakers (14).

Prospective study for population

Since we obtained hopeful findings at the beginning of our case-control study, and since there were no studies on the preventive effect of ginseng on cancer, we decided to perform a more reliable Cohort study in that year of 1987. This study was conducted in Kangwha-eup from August 1987 to December 1992. We studied 4,634 (2,362 men, 2,272 women) adults over 40 years old who completed a questionnaire on ginseng intake. Among 355 (7.7%) total deaths, subjects with

Table 3. Odds ratios for cancers and 95% confidence interval (CI) according to the type of ginseng

Type of ginseng		Cancer patients(%)	Controls (%)	OR	95% CI
No intake of ginseng		921	605	1.00	Reference
Intake of ginseng		1,066 (53.6)	1,382 (69.6)	0.50	0.44-0.58
Fresh ginseng					
	Fresh slice	210	172	0.79	0.63-1.01
	Juice	69	63	0.71	0.49-1.03
	Extract	146	255	0.37	0.29-0.46
	Extract & powder	15	38	0.40	0.24-0.66
	Extract & fresh slice	8	16	0.32	0.14-0.73
White ginseng					
	Extract	373	442	0.57	0.48-0.68
	Extract & fresh slice	68	79	0.55	0.39-0.77
	Extract & powder	15	41	0.22	0.12-0.38
	Powder	60	129	0.30	0.22-0.41
	Powder & fresh slice	21	31	0.39	0.22-0.67
	Tea	43	41	0.69	0.45-1.07
Red ginseng					
	Extract	6	17	0.20	0.08-0.50
Others		22	58	0.16	0.10-0.25

Adjusted for age, sex, marital status, education, smoking, alcohol consumption
Cancer Epidemiology, Biomarkers & Prevention (ref. 14)

Table 4. Odds ratios for cancers according to frequency and lifetime consumption of ginseng

Frequency of ginseng intake	Total				Male				Female			
	Case	Controls	OR*	95% CI	Cases	Controls	OR**	95%CI	Cases	Controls	OR**	95% CI
Frequency of ginseng intake None	921	605	1.00	Reference	409	234	1.00	Reference	512	371	1.00	Reference
1-3 times/year	417	440	0.60	0.51-0.71	246	231	0.62	0.49-0.79	171	209	0.60	0.47-0.76
4-11 times/year	324	394	0.51	0.43-0.61	197	223	0.48	0.37-0.62	127	171	0.54	0.42-0.71
1 time/month or more	325	548	0.36	0.30-0.43	220	384	0.31	0.25-0.39	105	164	0.47	0.35-0.62
Total lifetime consumption of ginseng												
None	921	605	1.00	Reference	409	452	1.00	Reference	512	371	1.00	Reference
1 - 50	774	903	0.55	0.47-0.63	452	501	0.51	0.42-0.63	322	402	0.58	0.48-0.71
51 - 100	103	139	0.46	0.35-0.61	75	100	0.41	0.29-0.58	28	39	0.56	0.34-0.93
101 - 300	109	185	0.34	0.27-0.45	80	131	0.32	0.23-0.44	29	54	0.39	0.25-0.61
301 - 500	28	50	0.31	0.19-0.49	20	29	0.33	0.18-0.62	8	21	0.29	0.14-0.63
501 -	52	105	0.28	0.20-0.39	36	77	0.25	0.16-0.38	16	28	0.42	0.23-0.79

* Adjusted for age, sex, marital status, education, smoking, alcohol consumption.** Adjusted for age, marital status, education, smoking, alcohol consumption.

Reproduced with permission from Cancer Epidemiology, Biomarkers & Prevention (ref.14)

cancer totalled 137 (3.0%) among studied subjects, with 58 (1.3%) alive and 79 (1.7%) deaths. Cancer accounted for 79 (22.8%) of the total deaths. Of 4,634 persons eligible for analysis, 70.5% (3,267) were ginseng intakers. Ginseng intakers had a decreased relative risk (RR;0.40, 95% CI=0.28-0.56) compared with non-intakers. The RR of cancer according to the kind of ginseng was 0.31 (95% CI=0.13-0.74) for fresh ginseng extract intakers. There was a decrease in risk with rising frequency of ginseng intake, showing a statistically significant dose-response relationship (Table 6). Newly diagnosed cancer cases have–been identified: 42 stomach, 24 lung, 14 liver and 57 at other sites. The RRs of ginseng intakers were 0.33 (95% CI=0.18-0.57) in gastric cancer and 0.30 (95% CI=0.14-0.65) in lung cancer. Among ginseng preparations, fresh ginseng extract intakers were significantly associated with the decreased risk of gastric cancer (RR; 0.33, 95% CI=0.12-0.88) (Table 7). These results strongly revealed that Panax ginseng C.A. Meyer has preventive effect against cancer (15).

Constituents of ginseng and possible mechanism

In knowledge that ginseng has been taken as tonic for a long time in Korea, however, we are attaching more importance in our study to confirming whether ginseng has as great an anticarcinogenic effect in human beings as it has in rodents rather than studying its active components.

The source of ginseng is the root of *Panax ginseng* (Araliaceae), which is a perennial plant, and commercially available ginseng is classified into fresh, white and red ginseng. White ginseng is made by peeling roots and drying them without steaming. Red ginseng is made by the steaming and drying of fresh ginseng, suggesting the occurrence of chemical transformation by heat. The presence of saponin in ginseng was first reported by Garriques in 1854, when he isolated a saponin component from American ginseng, Panax quinquefolium and named it "Panaquilon". Brekhman et al reported saponin as the active component of ginseng in 1957, and the Shibata and Tanaka group reported that ginseng saponin was triterpenoidal glycosides of dammarane type with glucose, arabinose, xylose or rhamnose, and named them ginsenoside-Rx as active components. Thirty-four kinds of ginsenosides have so far been isolated from fresh, white or red ginseng, among which 22 kinds of ginsenosides are protopanaxadiol type and 11 kinds of ginsenosides are protopanaxatirol type, and only ginsenoside Ro, oleanane type. Since ginsenosides are generally labile under acidic conditions, ordinary acidic hydrolysis is always accompanied by many side reactions such as cyclization of side chains, glycosyl elimination and epimerization of carbone-20 by SN1 reaction. Accordingly, the chemical transformations of secondary metabolites have appeared during the steaming process of preparing red ginseng. The unique components of

Table 5. Odds ratios for various cancers according to ginseng intakers

Site of cancer	Cases Never taken/ ever taken	Controls Never taken/ ever taken	OR	95% CI
Lip, oral cavity & pharynx	67/ 92	40/119	0.47	0.29-0.76
Esophagus	40/ 47	14/ 73	0.20	0.09-0.38
Stomach	142/158	76/224	0.36	0.25-0.52
Colon & rectum	55/ 63	32/ 86	0.42	0.24-0.74
Liver	108/156	67/197	0.48	0.33-0.70
Pancreas	12/ 11	5/ 18	0.22	0.05-0.95
Larynx	21/ 19	8/ 32	0.18	0.06-0.54
Lung	120/156	81/195	0.55	0.38-0.79
Female breast	87/ 92	70/109	0.63	0.40-1.05
Cervix uter	156/146	132/170	0.72	0.52-1.01
Ovary	17/ 5	8/ 14	0.15	0.04-0.60
Urinary bladder	23/ 40	16/ 47	0.64	0.28-1.47
Thyroid gland	16/ 24	14/ 26	0.96	0.38-2.44
Others	53/ 61	35/ 79	0.48	0.27-0.85

Adjusted for age, sex, marital status, education, smoking, alcohol consumption
Cancer Epidemiology,Biomarkers & Prevention (ref.14)

Table 6. Relative risks of cancer by ginseng Intake

Kinds of ginseng	Number of noncases	Number of cancer	Adjusted RR	95%CI
No intake	1,283	62	1.00	-
Frequency of ginseng intake				
1 - 3 times/year	1,439	39	0.46	0.30-0.69
4 - 11 times/year	924	21	0.35	0.21-0.58
12 times/month or more	804	15	0.34	0.20-0591
Total	4,450	137		

Adjusted for age, sex, education, smoking and alcohol consumptionRR=relative risk, CI=confidence interval
Intl. J. Epidemiol. (ref.15)

Table 7. Adjusted relative risks and 95 confidence interval for selected cancer by ginseng intake

Ginseng intake	Number of subjects	Stomach (42)			Lung (24)			Liver (14)		
		No.	RR	95% CI	No.	RR	95% CI	No.	RR	95% CI
No intake	1,283	23	1.00	-	14	1.00	-	4	1.00	-
Ginseng intake	3,167	19	0.33	0.18-0.57	10	0.30	0.14-0.65	10	0.86	0.25- 2.94
Fresh ginseng	236	2	0.57	0.17-1.94	1	0.67	0.15-3.43	2	1.97	0.34-2.95
Fresh ginseng extract	296	1	0.33	0.12-0.88	1	0.28	0.04-2.17	-	-	-
White ginseng powder	147	1	0.24	0.03-1.84	-	-	-	-	-	-
White ginseng extract	68	2	1.34	0.30-5.97	-	-	-	-	-	-
Boiled chicken with young ginseng root	381	5	0.43	0.12-1.43	1	0.80	0.08-1.95	1	0.85	0.15-4.87
Ginseng tea	442	6	0.64	0.26-1.61	4	0.35	0.26-2.44	2	1.72	0.36-8.26

Adjusted for age, sex, education, smoking and alcohol consumption
( ): Number of cancer
RR=relative risk, CI=confidence interval
Intl. J. Epidemiol. (ref.15)

17. Konoshima, T.; Kokumai, M.; Kozuka , M. et al. *Proceeding of the 38th Annual Meeting of the Japanese Society of Pharmacognosy* Japanese Society of Pharmacognosy, Kobe, Japan, 1991, 125.
18. Wu, X.G.; Zhu, D.H. *J. Tongji Med. Univ. China.* 1990, 10, 141-5.
19. Bespalov, V.G.; Aleksandrov, V.A.; Davydov, V.V.; Limanko, A.Yu ; Molokovskii, D.S.; Petrov, A.S.; Slepyan, L.I.; Trilis, Ya G. *Bulletin of Experimental Biology and Medicine* 115, 1993, 63-65.
20. Park, J.D. *Korean J. Ginseng Sci.* 1996, 20, 389-415.
21. Kenarova, B.; Neychev, H.; Haejiivanova, C.; Petkov, V.D.; *Jpn. J. Pharmacol.* 1990, 54, 447-54.
22. Odsashima, S.; Nakayabu, Y.; Honjo, N.; Abe, H.; Arichi, S. *Eur. J. Cancer* 1979, 15, 855-92.
23. Tode, T.; Kikuch, V.; Kita, T.; Hirata, J.; Imaizumi, E.; Nagata, I; *J. Cancer Res. Clin. Oncol.* 1993, 120, 24-6.
24. Lee, Y.N.; Lee, H.Y.; Chung, H.Y.; Kim, S.I.; Lee, S.K.; Park, B.C.; Kim, K.W. *Eur. J. Cancer* 1996, 32A, 1420-1428.
25. Matsunaga, H.; Katano, M.; Yamamoto, H.; Mori, M.; Takata, K. *Chem. Pharm. Bull.* 1989, 37, 1279-81.
26. Matsunaga, H.; Katano, M.; Yamamoto, H.; Fujito, H.; Mori, M.; Takada, K.; *Chem. Pharm. Bull.* 1990, 38, 3480-82.
27. Matsunaga H,; Saita, T.; Nagumo, T.; Mori, M.; Katano, M. *Cancer Chemother. Pharmacol.* 1995, 35, 291-296.
28. Kim, J.Y.; Germolec, D.R.; Luster, M.I. *Immunophamacol. Immunotoxicol.* 1990, 12, 257-76.
29. Tomoda, M.; Takeda, K.; Shimizu, N. et al. *Biol. Pharm. Bull.* 1993,16, 22-5.

Chapter 17

Possible Inhibition of Atherosclerosis by a Flavonoid Isolated from Young Green Barley Leaves

Takashi Miyake[1], Yashihide Hagiwara[1], Hideaki Hagiwara[2], and Takayuki Shibamoto[1,3]

[1]Department of Environmental Toxicology, University of California, Davis, CA 95616
[2]Hagiwara Institute of Health, 1173 Maruyama, Asazuma-cho, Kasai 679-01, Japan

Young green barley leaves are known to possess potent pharmacological properties, including antioxidative, anti-inflammatory, antimutagenic, and antiallergic activities. In particular, an flavonoid, 2"-O-glycosylisovitexin (2"-O-GIV), isolated from an ethanol extract of young green barley leaves, possesses a strong inhibitory effect toward lipid peroxidation. 2"-O-GIV inhibited acetaldehyde formation from LDL by 76% at a level of 1 µmol/50 µg, whereas ferulic acid inhibited by 66% at the same level. In a case of a blood plasma system, 2"-O-GIV and probucol inhibited acetaldehyde formation by 89% and 94%, respectively, at a level of 3 µmol. 2"-O-GIV and vitamin C inhibited MDA formation by 54% and 32%, respectively, at a level of 0.1 µmol. A synergetic effect between 2"-O-GIV and vitamin C was observed.

Barley has been cultivated and fed to livestock since ancient times. Essence extracted from young green barley leaves has been reported to exhibit many biological characteristics including anti-aging, anti-carcinogenesis, anti-diabetic, and anti-arteriosclerosis (1). However, no scientific proof of these characteristics existed until a potent anti-oxidant was isolated and identified in a green barley leaf essence (2). This novel natural antioxidant, which is a flavonoid, was identified as 2"-O-glycosyl isovitexin (2"-O-GIV, Figure 1). Since the discovery of this flavonoid, the antioxidative activities of 2"-O-GIV examined in various lipid peroxidation systems including squalene/UV (3), ethyl ester of fatty acids/Fenton's reagent (4), phospholipids or cod liver oil/Fenton's reagent (5), and ω-3 fatty acids/Fenton's reagent have been reported(6).

Lipid peroxidation model systems have been used most commonly to investigate biological activities of chemicals because lipid peroxidation is associated with many biological complications such as carcinogenesis, mutagenesis, aging, and atherosclerosis (7-9) as well as with human immunodeficiency virus (HIV) progression (10). However, its mode of toxic action is not yet clearly understood. Lipids produce many low molecular weight carbonyl compounds upon oxidation (11). Therefore, these carbonyl compounds such as malondialdehyde (MDA), glyoxal, acrolein, acetaldehyde, and formaldehyde which directly crosslink to proteins and bind

[3]Corresponding author.

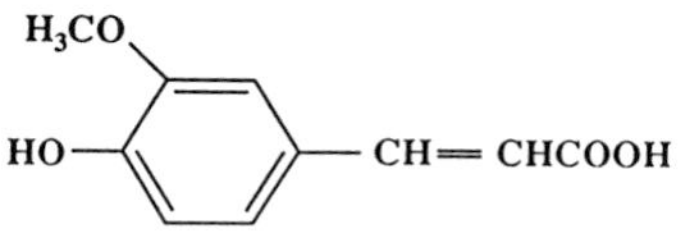

Ferulic acid

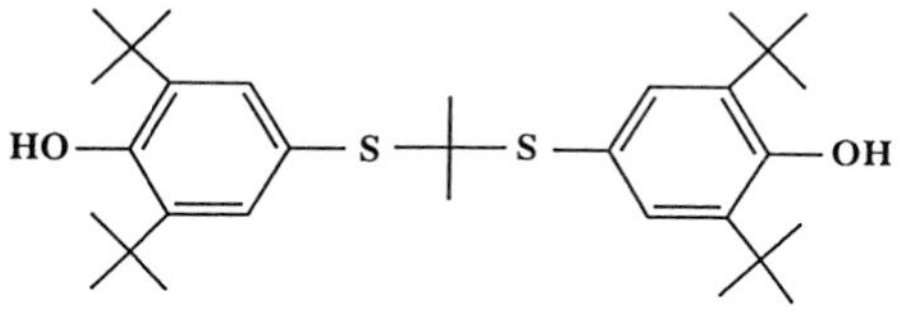

Probucol

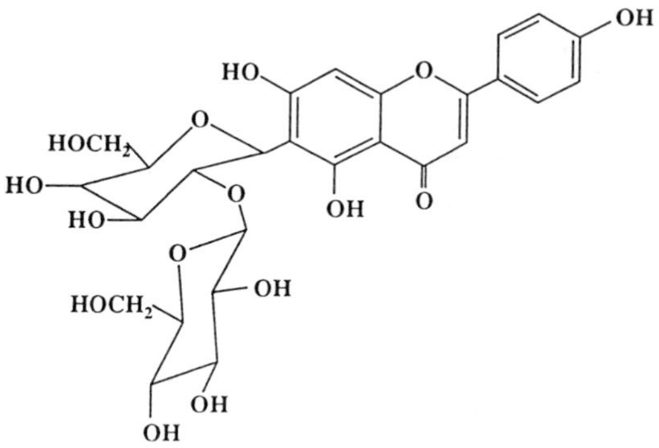

2"-O-Glycosylisovitexin (2"-O-GIV)

Figure 1. Structure of ferulic acid, probucol, and 2"-O-glycosyl isovitexin.

covalently to nucleic acids (12) may possibly play an important role in the toxic effects caused by lipid peroxidation (13, 14).

In addition to initiating these adverse effects, low molecular weight aldehydes formed from lipid peroxidation can be used as indicators to detect oxidation in lipids. Therefore, many studies have been conducted using these aldehydes as indicator, in particular, malondialdehyde (MDA). The formation of acetaldehyde was also used to monitor oxidative reaction mechanisms of L-ascorbic acid (15).

In the present study, the antioxidative activity of 2"-O-GIV was examined using low density lipoprotein and blood plasma oxidized with Fenton's reagent.

Experimental

Chemicals. L Ascorbic acid (reagent grade), butylated hydroxytoluene (BHT), Trizma® hydrochloride, Trizma® base, fatty-acid-free bovine serum albumin, probucol, and fat red 7B were purchased from Sigma Chemical Co. (St. Louis, MO). Cysteamine hydrochloride, 2,4,5-trimethylthiazole, ferulic acid, sodium dodecyl sulfate (SDS), hydrogen peroxide, and ferrous chloride were obtained from Aldrich Chemical Co. (Milwaukee, WI). Hydrogen peroxide was obtained from Fisher Scientific Co., Ltd. (Fair Lawn, NJ). The standard stock solution of 2,4,5-trimethylthiazole was prepared by adding 10 mg of 2,4,5-trimethylthiazole to 1 mL of dichloromethane and was stored at 5 °C. Authentic 2-methylthiazolidine was synthesized according to the method reported previously (16). The structures of probucol and ferulic acid are shown in Figure 1.

2"-O-Glycosylisovitexin (2"-O-GIV) was isolated from young green barley leaves (*Hordium vulgare L. var. nudum Hook*) harvested two weeks after germination by a method previously reported (2) using column chromatography with Amberlite XAD-2 nonionic polymeric absorbent. The structure of 2"-O-GIV is shown in Figure 1.

Preparation of Low-Density Lipoprotein (LDL). LDL was prepared from blood sample obtained from a male quarter horse (5 yeare old) according to the method reported previously (15). After filtration sterilization (0.45 μm; Nalge Stbron) of the LDL, the protein concentration was determined by the Coomassie Blue dye-binding assay (17). A 10-μL aliquot of appropriately diluted LDL was added to the dye reagent, the solution mixed, and absorbance at 594 nm measured versus a reagent blank, using a Hewlett-Packard 8452A diode array UV spectrophotometer. A standard curve using bovine serum albumin was used to calculate the LDL concentration.

Preparation of Blood Plasma. The blood from a male quarter horse (5 years old) was collected in a sterile, 3.5-mL tube containing 60 μL of a 7.5% EDTA solution (4.5 mg EDTA). Plasma and red blood cells were separated following centrifugation (5000 rpm for 30 min at 4 °C) and immediately frozen on dry ice. The plasma was stored at - 80 °C until use.

Oxidation of LDL and Blood Plasma With Fenton's Reagent With or Without Antioxidants. A 5-mL aqueous solution containing 23 μL of LDL (final concentration, 2.17 μg/mL), 0.25 mmol of Trizma® buffer (pH 7.4), 0.75 mmol of potassium chloride, 1 mmol of ferrous chloride, and 0.5 mmol of hydrogen peroxide was incubated with 2"-O-GIV (amounts ranging from 0 to 1 μmol, with 0.1

amount of cholesterol esters which may be associated with the development of atherosclerosis. Figure 4 shows the inhibitory effect of 2"-O-GIV and ferulic acid against LDL oxidation. Acetaldehyde was used to monitor the formation and inhibition of lipid peroxidation because MDA tends to be trapped with proteins. Over 2 nmol of acetaldehyde was formed from 50 μg of LDL. Formation of acetaldehyde decreased when the amount of either 2"-O-GIV or ferulic acid was increased. Ferulic acid inhibited acetaldehyde formation by 50% at the level of 0.3 μmol/50 μg of LDL) whereas 2"-O-GIV required 0.7 μmol/50μg of LDL to obtain the same level of inhibition. On the other hand, 2"-O-GIV inhibited acetaldehyde formation by 76% at the level of 1 μmol/50 μg whereas ferulic acid inhibited by 66% at the same level.

Figure 5 shows the antioxidative activities of 2"-O-GIV and probucol measured in a blood plasma system. Probucol was used to examine the relative antioxidative activity of 2"-O-GIV because it is a commercial product with a million dollar sales in Japan and has been used to treat atherosclerosis. However, probucol produced a significant amount of MDA by oxidation (T. Miyake and T. Shibamoto, unpublished data). Therefore, acetaldehyde was used as an indicator of oxidation of blood plasma. The antioxidative activities of 2"-O-GIV and probucol were almost identical. When blood plasma (516 μg) was oxidized without an antioxidant, 135 nmol of acetaldehyde was recovered. Inhibitory activity of both 2"-O-GIV and probucol toward acetaldehyde formation increased greatly when their levels increased over 0.7 μmol. 2"-O-GIV and probucol inhibited acetaldehyde formation by 89% and 94%, respectively, at the level of 0.3 μmol). The results indicate that 2"-O-GIV may inhibit atherosclerosis.

Figure 6 shows the antioxidative activities of 2"-O-GIV and vitamin C in a blood plasma system. When blood plasma was oxidized with Fenton's reagent, 1.11 nmol/μg (blood plasma) of MDA was recovered. However, no MDA was recovered from unoxidized blood plasma. In this experiment, MDA was used as an indicator of oxidation because vitamin C itself produced a significant amount of acetaldehyde by oxidation with Fenton's reagent (15). 2"-O-GIV and vitamin C inhibited MDA formation by 54% and 32%, respectively, at the level of 0.1 μmol. On the other hand, when equal mols of 2"-O-GIV and vitamin C were mixed, a 75% inhibitory effect was obtained at the level of 0.1 μmol (total). A synergetic effect between 2"-O-GIV and vitamin C was observed. The antioxidative activities of flavonoid such as 2"-O-GIV are due to their ability to chelate metal ions (such as Fe^{2+}) by means of either the 3-hydroxy, 4-keto grouping or the 5-hydroxy, 4-keto grouping, in addition to scavenging free radicals deriving from the phenolic moiety of the structure (27).

Conclusion

Atherosclerosis is one of the most common diseases in developed countries, such as Japan and the U.S., where people tend to eat more fatty foods. Detail mechanisms of atheroscrelosis are not yet clearly understood but there is a strong evidence that lipid peroxidation plays an important role in the initiation of atheroscrelosis. Therefore, antioxidants such as probucol have been used to treat atheroscrelosis. However, use of synthetic antioxidants (e.g., butylated hydroxy toluene) in human foods has been restricted because of their possible chronic toxicity. 2"-O-GIV is a natural plant product and large amounts of barley leaves (which contain 0.5-0.7% of 2"-O-GIV) have been consumed by livestock for many years without any evidence of adverse effects. 2"-O-GIV may be useful in the treatment of atherosclerosis. Therefore, further investigation of its physiological effects should be undertaken.

184

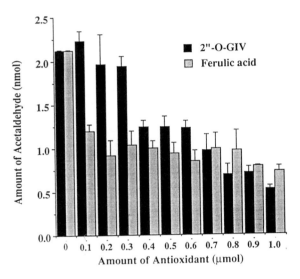

Figure 4. Inhibitory effect of 2"-O-GIV and ferulic acid on LDL oxidation.

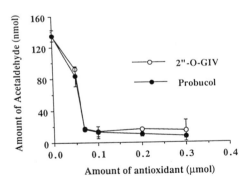

Figure 5. Antioxidative activities of 2"-O-GIV and probucol measured in blood plasma.

red ginseng are known as 20(R)-ginsenosides Rh1, and Rh2, 20(S)-ginsenoside Rh2, 20(R)-ginsenoside Rg2, 20(S)-ginsenoside Rg3, and notoginsenoside R4. Malonyl-ginsenoside-Rb1, -Rb2, -Rc, and Rd, found only in white ginseng, are transformed into ginsenosides Rb1, Rb2, Rc, and Rd, ginsenoside -Rs1, or Rs2, saponin characteristics of red ginseng by decarboxylation. Among chemical constituents other than saponin, 1-2% ether soluble components are contained in the root of ginseng. Twelve kinds of phenolic compounds, including salicylic acid, caffeic acid, and maltol, which shows antioxident activity, are isolated from ginseng. Maltol especially, which is only present in red ginseng and produced from maltose by amino-carbonyl reaction, shows antioxidant activity. Nine kinds of polyacetylene compounds have been isolated and elucidated as pananaxynol, panaxydol, panaxytriol, acetylpanaxydol, chloropapaxydol, panaxyne, and so on, and also ginsenoyne A, B, C, D, E, F, G, H, I, J, K have been reported from hexane soluble fraction. As regards essential oils, about 30 kinds of sesquiterpenes including azulene and patchoulene have been identified from the ether soluble fraction of fresh ginseng, and five kinds of methoxypyrazine and eight kinds of alkypyrazine derivatives have been identified from the basic fraction of the ether soluble extract. Sesquiterpene alcohol, panasinsanols A and B have also been isolated. Seven kinds of β-caboline alkaloid have been isolated from the ether soluble alkaloidal fraction and choline, has been isolated from the water soluble fraction. Twenty-one kinds of neutral or acidic polysaccharides which make up 50-60% of the root of ginseng, are purified and named panaxan A-U consisting of glucose, arabinose, galatose, rhamnose, xylose or uronic acid. Ginseng contains about 12 to 15 % of nitrogen containing compounds, which are comprised of amino acids and oligopeptides such as adenosine and pyroglutamic acid, and Arg-Fru-Glc is formed during the preparation of red ginseng by amino-carbonyl reaction. Other vitamins, inorganic substances, free monosaccharides, and organic acids are also contained in ginseng (20).

Considered to be the active components of ginseng, biomass of ginseng tincture have reported for their antimutagenic activity (16). Ginsenoside Rg1 induced an augmentation of the production of IL-1 by macrophages and exerted a direct mitogenic effect on microcultured thymus cells (21). Morris hepatoma cells (Mh1C1) were reversely transformed by ginsenosides (oligoglycosides) with 20(S)-protopanaxadiol and 20(S)-protopanaxatirol as aglycones, showed much less ability to grow in culture (22). Inhibitory effects by oral administration of ginsenoside Rh2 on the growth of human ovarian cancer cells in nude mice were reported (23). The ginsenosides Rh1 and Rh2 caused the differentiation of G9 teratocarcinoma cells (24). A polyacetylenic alcohol, panaxytriol isolated from *Panax ginseng* C.A. Meyer inhibited both tumor cell growth in vitro and the growth of B16 melanoma transplanted mice, and panaxytriol rapidly inhibited cellular respiration and disrupted cellular energy balance in breast carcinoma cells (Breast M25-SF) (25-

27). The acidic polysaccharide from the root of *Panax ginseng* C.A. Meyer showed remarkable reticuloendothelial system-potentiating activity, pronounced anti-complementary activity and alkaline phosphatase-inducing activity in a dose dependent manner (28, 29). Judging from the fact that each constituent has its own characteristics, one or more than one in cooperation seem to reduce cancer risks. It is, of course , possible that some unknown constituents are related to the reduction of cancer risks.

As a conclusion from the results of the above experimental and epidemiological studies, we have shown that *Panax ginseng* C. A. Meyer had varied anticarcinogenic effects depend on type and age of ginseng and preventive effects on various human cancers and that it contained some components of anticarcinogenesis and cancer prevention. We think ginseng should be recognized as a functional food for cancer prevention, and that further studies for the identification of its active components, mechanisms of action and clinical interventions should be undertaken with world-wide collaboration.

Literature Cited

1. Goodman, L.S.; Wintrobe, M.M.; Dameshek, W.; Goodman, M.J.; Gilman, A.; McLennan, M.T. *J.A.M.A.* 1946, 132, 126.
2. Beardsley, T. *Scientific American* 1994, January, 130-138.
3. American Cancer Society: Cancer Facts & Figures-1996, American Cancer Society Inc., Atlanta, GA, U.S.A.
4. Jing, T.H.; Jing, S.B.A. *Simplified Version of Shennong's Ancient Chinese Medical Book* Liang Dynasty of China, circa 500 A.D., Munkwang Doso, Taipei, 1982, p40.
5. Yun, T.K.; Yun, Y.S.; Han, L.W. *Proc. 3rd International Ginseng Symp.* Korea Ginseng Research Institute Press, Seoul, 1980, 87-113.
6. Yun, T.K.; Yun,Y.S.; Han, I.H. *Cancer Detect. Prev.* 1983, 6, 515-25.
7. Yun, T.K.; Kim, S.H.; Oh, Y.R. *J. Korean Cancer Assoc.* 1987, 19, 1-7.
8. Yun,T.K.; Kim, S.H. *J. Korean Cancer Assoc.* 1988, 20, 133-42.
9. Yun, T.K. *J.Toxicol. Sci.(Japan)* 16: Suppl.1, 1992, 53-62.
10. Yun, T.K.; Kim, S.H.; Lee, Y.S. *Anticancer Res.* 1995, 15, 839-46.
11. Yun, T.K.; Lee, Y.S. *Korean J. Ginseng Sci.* 1994, 18, 89-94.
12. Yun, T.K.; Lee, Y.S. *Korean J. Ginseng Sci.* 1994, 18, 160-4.
13. Yun, T.K.; Choi, S.Y. *Intl. J. Epidemiology* 1990, 19, 871-6.
14. Yun, T.K.; Choi, S.Y. *Cancer Epidemiol. Biomarkers Prev.* 1995, 4, 401-8.
15. Yun,T.K.; Choi, S.Y. *Intl. J. Epidemiol.* 1998, 27, in press.
16. Umnova, N.V.; Michurina, T.L.; Sminova, N.L.; Aleksandrova, I.V.; Poroshenko, G.G. *Bull. Exp. Biol. Med.* 1991, 111, 507-9.

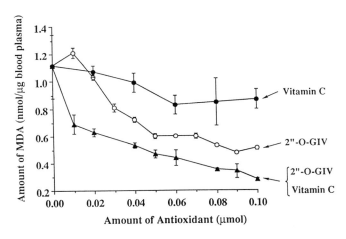

Figure 6. Antioxidative activities of 2"-O-GIV and vitamin C in blood plasma.

References

1. Hagiwara, Y. In *A Good Health Guide*. Passwater, R. A., Ed.; Keats Publishing, Inc.: New Canaan, CT, 1985.
2. Osawa, T.; Katsuzaki, H.; Hagiwara, Y.; Hagiwara, H.; Shibamoto, T. J. Agric. Food Chem. **1992**, *40*, 1135–1138.
3. Nishiyama, T.; Hagiwara, Y.; Hagiwara, H.; Shibamoto, T. J. Am. Oil Chem. Soc. **1993**, *70*, 811–813.
4. Nishiyama, T.; Hagiwara, Y.; Hagiwara, H.; Shibamoto, T. J. Agric. Food Chem. **1994**, *42*, 1728–1731.
5. Nishiyama, T.; Hagiwara, Y.; Hagiwara, H.; Shibamoto, T. Food Chem. Toxicol. **1994**, *32*, 1047–1051.
6. Ogata, J.; Hagiwara, Y.; Hagiwara, H.; Shibamoto, T. J. Am. Oil Chem. Soc. **1996**, *73*, 653–656.
7. Harman, D. In *Free Radicals in Biology*. Pryor, W. A., Ed.; Academic Press: Orlando, FL, Vol 5. 1982; Chapter 8.
8. Cutler, R. G. In *Radicals in Biology*. Pryor, W. A. Ed.; Academic Press: Orlando, Vol. 6. 1984; Chapter 11
9. Pryor, W. A. In *Antimutagenesis and Anticarcinogenesis Mechanisms*. Shankel, D. M., Hartman, P. E., Kada. T., Eds.; Plenum Press: New York, 1986; pp 45–59.
10. Baruchel, S.; Wainberg, M. A. J. Leukocyte Biol. **1992**, *52*, 111–114
11. Esterbauer, H. In *Free Radicals*. McBrien, D. C. H., Slater, T. F., Eds.; Academic Press: London, 1982; pp 101–128.
12. Lam, C. W.; Casanova, M.; Heck, H. D. Fundam. Appl. Toxicol. **1986**, *6*, 541–550.
13. Basu, A. K.; Marnett, L. J. Carcinogenesis **1983**, *4*, 331–333.
14. Shamberger, R.J.; Andreone, T.L.; Willis, C. E. J. Natl. Cancer Inst. **1974**, *53*, 1771–1773.
15. Miyake, T.; Shibamoto, T. J. Agric. Food Chem. **1995**, *43*, 1669–1672.
16. Yasuhara, A.; Shibamoto, T. Agric. Biol. Chem. **1989**, *53*, 2273–2274.
17. Bradford, M. M. Anal. Biochem. **1976**, *72*, 248–254.
18. Miyake, T; Shibamoto, T. J. Agric. Food Chem. **1993**, *41*, 1968–1970.
19. Wong, J. W.; Ebeler, S. E.; Rivkah-Isseroff, R.; Shibamoto, T. Anal. Biochem. **1994**, *220*, 73-81.
20. Ames, B. N.; Shigenaga, M. K.; Hagen, T. M. Proc. Natl. Acad. Sci. U.S.A. **1993**, *90*, 7915–7922.
21. Belch, J. J. F. Scottish Med. J. **1992**, *37*, 67–68.
22. Retsky, K. L.; Freeman, M. W.; Frei, B. J. Biol. Chem. **1993**, *268*, 1304–1309.
23. Glavind, J.; Hartmann, S.; Clemmesen, J.; Jessen, K. E.; Dam, H. Acta Pathol. Microbiol. Scand. **1952**, *30*, 1–6.
24. Aoyama, S.; Iwakami, M. Jpn. Heart J. **1965**, *6*, 128–143.
25. Jialal, I.; Grundy, S. M. J. Clin. Invest. **1991**, *87*, 597–601.
26. Jensen, R. G. In *Fatty Acids in Foods and Their Health Implications*. Chow, C. K., Ed.; Marcel Dekker: New York, 1992; p 129.
27. Pratt, D. E. In *Phenolic, Sulfur and Nitrogen Compounds in Food Flavors*. Charalambous G, Kats I.; Eds.; ACS Symposium Series 26, American Chemical Society: Washington, DC, 1976, pp 1-13.

MEASUREMENT OF FUNCTIONAL SUBSTANCES

Chapter 18

The Genox Oxidative Stress Profile

An Overview on Its Assessment and Application

Richard Cutler[1], Kristine Crawford[1], Roy Cutler[1], Rong-Zhu Cheng[1,2], Narasimhan Ramarathnam[1,2], Masao Takeuchi[2], and Hirotomo Ochi[2]

**[1]Genox Corporation, 1414 Key Highway, Baltimore, MD 21230
[2]Japan Institute for the Control of Aging, 723-1 Haruoka, Fukuroi, Shizuoka 437-01, Japan**

A growing and compelling body of scientific evidence now exists indicating oxidative stress (OS) as a major causative factor in many age-related dysfunctions and specific diseases. Since the oxidative stress state (OSS) of an individual depends on hereditary, dietary, and environmental factors, there is a large heterogeneity in the population that may be related to disease incidence and longevity. Hence there is a need to assess how well an individual is coping against OS. The Genox Corporation, with the assistance of the Japan Institute for the Control of Aging has developed a new medical diagnostic and research technique called "Oxidative Stress Profiling (OSP)". The Genox OSP consists of about 44 different assays measuring the levels of oxidative damage in lipids, proteins, and nucleic acids, and the antioxidant defenses in the serum. A general overview is presented on the Genox OSP and its application in the testing of human subjects.

Human beings maintain their lives by obtaining energy through respiration. Once inhaled into the body, triplet oxygen ($^3O_2$) is utilized in energy metabolism, and is finally reduced to water. In this process, a small portion of it is converted into reactive oxygen species (ROS) such as the superoxide radical (O_2^-), hydrogen peroxide (H_2O_2), and the hydroxyl radical ($^.OH$) (1). In addition, several external factors such as ultraviolet light, radiation, asbestos, heavy metal ions, carcinogens, pollution, and cigarette smoke also induce the generation of free radicals (2-8). The free radicals thus generated are involved in various *in-vivo* oxidation reactions and alter a wide range of physiological functions and metabolic components (9).

Despite the presence of various inherent defense systems such as enzymic (catalase, superoxide dismutase, glutathione peroxidase) and nonenzymic (ascorbate, tocopherol, etc.) antioxidants, overwhelming levels of active oxygen radicals are known to react with DNA, proteins, and lipids and cause damage to these compository components of the body (*10-11*). This process leads to the loss of enzyme activity, decrease in the solubility of proteins, and result in the fragmentation of the chromosomes (*12*).

It is also known that DNA can receive multiple types of damages such as oxidation of pyrimidine or purine bases in the nucleoside, and the oxidation of the sugar moieties (*13*). A comparison of the amount of oxygen consumption of individual mammalian species and their life spans has shown that the animals with greater oxygen consumption per antioxidant level have relatively shortened life span (*14*). Also, an analysis of the urine collected from animals with greater amount of oxygen consumption per antioxidant level showed higher levels of 8-hydroxy-2'-deoxyguanosine (8-OHdG) excretion into the urine.

Free radicals, oxidative stress, and the aging process

All biological constituents of any organism are the result of "trade-offs" of benefits versus disadvantages that have taken place under millions of years of evolutionary pressure. The human aging process itself appears to be a result of such trade-offs involving the side reactions of development and energy metabolism.

A compelling body of scientific evidence now indicates that many dysfunctions and diseases in humans are products of such a trade-off called *Oxidative Stress*. Examples of oxidative stress related diseases are outlined in Figure 1.

Indeed, the general age-dependent decline in optimum health and performance (known as *Normal Aging*) appears to be the result of ROS, which is one of its major causative factors. In addition, many different types of bacterial, fungal, and viral infections increase the amount of ROS generated *in vivo*. Sometimes this increase is dramatic (as in AIDS), but it also can be very slight as in a number of low-grade bacterial and fungal infectious diseases.

Oxidative Stress (OS) is defined as the steady-state level of oxidative damage within a cell, tissue or organism caused by ROS. The degree of Oxidative Stress or the **Oxidative Stress State (OSS)** present in a given biological system is determined by the net result of three major factors - **(1)** initial rate of generation of ROS, **(2)** level of antioxidant protective processes, and **(3)** the rate of repair and general turnover or removal rate of the oxidized targets that include nucleic acids, proteins, and lipids. A scheme to illustrate the Oxidative Stress State is shown in Figure 2. Many of the oxidized damage components that are produced throughout the body are transported to the serum, urine or breath as illustrated **(4)**. The OSS, as shown in **(5),** can refer to any component or system such as the whole individual, an organ, a tissue, a cell or a sub-cellular fraction.

It is the ratio of **damage input (6)** to **damage output (7)** that determines the OSS value. This ratio is largely controlled by a specific set of genes known as "longevity determinant genes." The concept of OSS is fundamental to our understanding of health maintenance because it determines the probability that initiation of abnormal functions and disease will occur over time **(8).**

Aging

- *Normal aging processes,
  but at a higher than normal rate*
- *Segmental progeria disorders
  (Down's syndrome)*

Heart and Cardiovascular Disease

- *Atherosclerosis*
- *Adriamycin cardiotoxicity*
- *Alcohol cardiomyopathy*

Kidney

- *Autoimmune nephritic syndromes*
- *Heavy metal nephrotoxicity*
- *Solar radiation*
- *Thermal injury*
- *Porphyria*

Nervous System Disorders

- *Hyperbaric oxygen*
- *Parkinson's disease*
- *Neuronal ceroid lipofuscinoses*
- *Alzheimer's disease*
- *Muscular dystrophy*
- *Multiple sclerosis*

Red Blood Cells

- *Malaria*
- *Sickle cell anemia*
- *Fanconi's anemia*
- *Hemolytic anemia of prematurity*

Other Oxidative Stress Disorders

- *AIDS*
- *Radiation-induced injuries
  (accidental and radiotherapy)*
- *General low-grade inflammatory disorders*

Gastrointestinal Tract

- *Inflammatory and immune injury*
- *Diabetes*
- *Pancreatitis*
- *Halogenated hydrocarbon liver
  injury*

Eye

- *Cataractogenesis*
- *Degenerative retinal damage*
- *Macular degeneration*

Lung

- *Lung cancer (cigarette smoke)*
- *Emphysema*
- *Oxidant pollutants (O_3, NO_2)*
- *Bronchopulmonary dysphasia*
- *Asbestos carcinogenicity*

Iron Overload

- *Idiopathic hemochromatosis*
- *Dietary iron overload*
- *Thalassemia*

Ischemia Reflow States

- *Stroke*

Inflammatory-Immune Injury

- *Glomerulonephritis*
- *Autoimmune disease*
- *Rheumatoid arthritis*

Liver

- *Alcohol-induced pathology*
- *Alcohol-induced iron
  overload injury*
- *Organ transplantation*
- *Inflamed rheumatoid joints*
- *Arrhythmias*
- *Myocardial infarction*

Figure 1. Examples of oxidative stress related diseases in humans.

191

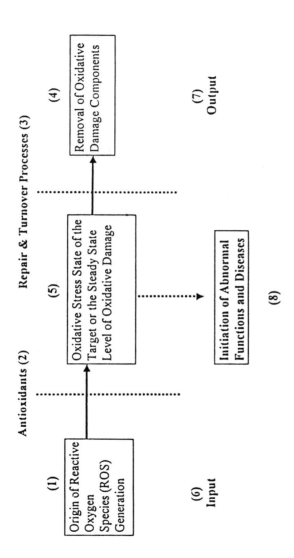

Figure 2. Scheme for the illustration of Oxidative Stress State.

Since the initiation and rate of progression with age of major diseases is strongly related to an individual's characteristic OSS, control of OSS is key to the control of human health and longevity. To further examine this hypothesis, there is a growing need within the scientific and medical communities for the development of specific, reliable, non-invasive, and cost-effective assays that are effective in measuring small changes of OSS. Further, the development of a unique, integrated set of these assays that can be used to calculate most effectively an individual's OSS is also essential.

The Genox Oxidative Stress Profile

The Genox Corporation seeks to meet these needs by offering an **Oxidative Stress Profile (OSP).** The **Genox OSP** provides the most complete set of assays designed to assess the oxidative damage and the antioxidant protection levels that determine the OSS of an individual. A typical Genox Oxidative Stress Profile (Genox OSP) is determined by an assortment of 44 different assays as illustrated in Figure 3.

The whole OSP is built upon seven different components, each contributing a specialized set of information towards the overall oxidative stress state of an individual. **Oxidative damage** is measured both in the serum and urine, and is determined by the levels of aqueous and lipid hydroperoxides, and total alkenals in the serum. Alkenals are highly reactive, and are known to react with proteins, enzymes, nucleic acids etc., causing alterations in their activity. The bio-markers of oxidative damage components in the urine are mainly 8-OHdG, total alkenals, and 8-epi-prostaglandin2α.

Levels of components in serum, such as free iron, ferritin, copper, and ceruloplasmin indicate the **pro-oxidant potential** of the serum. As bio-markers of diabetes, levels of **glucose** and **glycated protein** are measured. The total antioxidant capacity of the serum is measured by the **Oxygen Radical Absorption Capacity** (ORAC) assay, and the **Lipid Peroxidation Inhibition Capacity** (LPIC) assay that have been indigenously developed by us.

Levels of **specific antioxidants** in the serum, both aqueous (ascorbates, thiols, uric acid, and bilirubin) and lipid soluble, are measured to give a detailed information on what component is lacking in the patient. The lipid soluble antioxidants analyzed are the carotenoids (lutein, zeaxanthin, beta-cryptoxanthin, lycopene, alpha-carotene, beta-carotene, and retinol), and the tocopherols (alpha-, gamma-, and delta-tocopherols). The levels of coenzyme-Q10 are also reported, in addition.

Levels of serum lipids and proteins are good indicators of the general human health status. Hence, the information on the levels of cholesterol and triglycerides, bio-markers of cardiovascular-disease risk factors, is also equally important for the overall information generated from the OSP. Levels of albumin and globulin and the ratio of albumin/globulin that indicate the capacity of the body to resist diseases are the added features to the profile.

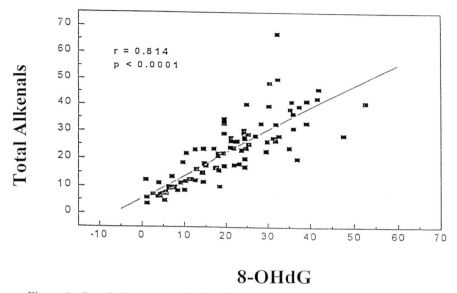

Figure 4. Correlation between the levels of total alkenals and the 8-OHdG levels in urine samples

Table I. Effect of Green Tea Drinking on the Urinary 8-OHdG Levels

Age Group (n)	Urinary 8-OHdG Levels During Control Period (μg)	Urinary 8-OHdG Levels During Green-Tea Intake Period (μg)
Total (n = 48)	16.2	14.6
60 and above (n = 4)	12.0	8.1
50 and above (n = 11)	12.3	12.5
40 and above (n = 12)	14.1	12.9
30 and above (n = 7)	24.3	19.9
20 and above (n = 14)	18.6	17.3
Male (n = 31)	17.5	16.4
Female (n = 17)	13.7	11.2

effectively the corrective action taken (such as dietary modification) is working or if for some reason beyond the individual's control, the OSP has changed over the years.

The OSP developed by the Genox Corporation is to meet these needs by providing effective and low-cost assays for measuring an individual's OSS causative and protective processes, using non-invasive methods of analysis. The OSP is designed for use by the scientific community in basic research, practicing clinicians and medical doctors, as well as individuals interested in personally optimizing their health and longevity.

Literature Cited

1. Ames, B. N.; Shigenaga, M. K. Oxidants are a major contributor to aging. *Ann. New York Acad. Sci.* **1992**, *663*, 85-96.

2. Dizdaroglu, M. Formation of an 8-hydroxyguanine moiety in deoxyribonucleic acid on γ-irradiation in aqueous solution. *Biochemistry* **1985**, *24*, 4476-4481.

3. Fiala, E. S.; Conaway, C. C.; Mathis, J. E. Oxidative DNA and RNA damage in the livers of Sprague-Dawley rats treated with the hepatocarcinogen 2-nitro- propane. *Cancer Res.* **1989**, *49*, 5518-5522.

4. Kasai, H.; Nishimura, S.; Kurokawa, Y.; Hayashi, Y. Oral administration of the renal carcinogen, potassium bromate, specifically produces 8-hydroxy-deoxyguanosine in rat target organ DNA. *Carcinogenesis* **1987**, *8*, 1959-1961.

5. Kasai, H.; Tanooka, H.; Nishimura, S. Formation of 8-hydroxyguanine resi-dues in DNA by X-irradiation. *Gann* **1984**, *75*, 1037-1039.

6. Kasai, H.; Nishimura, S. DNA damage induced by asbestos in the presence of hydrogen peroxide. *Gann* **1984**, *75*, 841-844.

7. Nair, U. J.; Floyd, R. A.; Nair, J.; Bussachini, V.; Friesen, M.; Bartsch, H. Formation of reactive oxygen species and of 8-hydroxydeoxyguanosine in DNA *in vitro* with betel quid ingredients. *Chem. Biol. Interact.* **1987**, *63*, 157-169.

8. Kasprzak, K. S.; Hernandez, L. Enhancement of hydroxylation and deglycosylation of 2'-deoxyguanine by carcinogenic nickel compounds. *Cancer Res.* **1989**, *49*, 5964-5968.

9. Kehrer, J. P. Free radicals as mediators of tissue injury and disease. *Crit. Rev. Toxicol.*, **1993**, *23*, 21-48.

10. Ursini, F.; Maiorino, M.; Sevanian, A. In *Membrane hydroperoxides.* Sies, H., Ed.; Oxidative Stress, Oxidants, and Antioxidants. Academic Press: San Diego; CA, 1991; p.319.

11. Stadtman, E. R. Metal ion catalyzed oxidation of proteins: biochemical mechanisms and biological consequences. *Free Rad. Biol. Med.* **1990**, *9*, 315-320.

12. Stadtman, E. R. Protein oxidation and aging. *Science* **1992**, *257*, 1220-1224.

13. Ames, B. N.; Shigenaga, M. K.; Hagen, T. M. Oxidants, antioxidants, and the degenerative diseases of aging. *Proc. Natl. Acad. Sci. USA* **1993**, *90*, 7915-7922.

14. Cutler, R. G. In *Antioxidants, aging and longevity.* Pryor, W. A., Ed.; Free Radicals in Biology, Academic Press: New York, NY, 1984, Vol. 6; 371-428.

15. Toyokuni, S.; Tanaka, T.; Hattori, Y.; Nishiyama, Y.; Yoshida, A.; Uchida, K.; Hiai, H.; Ochi, H.; Osawa, T. Quantitative immunohistochemical determination of 8-hydroxy-2'-deoxyguanosine by a monoclonal antibody N45.1: its application to ferric nitriloacetate induced renal carcinogenesis model. *Lab. Invest.* **1997,** *76*, 365-374.

Chapter 19

Pharmacokinetics of Soy Phytoestrogens in Humans

Shaw Watanabe[1] and Herman Adlercreutz[2]

[1]Department of Nutritional Science, Tokyo University of Agriculture,
Tokyo 156, Japan
[2]Department of Clinical Chemistry, The University of Helsinki, Helsinki, Finland

Phytoestrogens are potent food factors for the prevention of estrogen-related diseases. Determination of their absorption rate, metabolic rate, and excretion rate is necessary to know the proper dose to avoid toxic effects. A pharmacokinetic study on soybean isoflavonoids, after ingestion of 103 and 112 moles of daidzein and genistein, respectively, from kinako (soybean flour) was carried out in 7 healthy males. The peak plasma level of genistein (2,440±651 nmol/L) and daidzein (1,560±337 nmol/L) occurred 6 hours later after ingestion of kinako. Excretion of isoflavonoids in urine followed with the increasing concentration of plasma isoflavonoids. Daidzein was the main component in urinary excretion, but o-desmethylangolensin (O-DMA) and equol overwhelmed daidzein in the subjects who metabolized daidzein to them. Urinary excretion of isoflavonoids showed two or three peaks over 3 days. Fecal isoflavonoids often peaked again, even after defecation of the kinako meal.
        These findings suggest the presence of enterohepatic circulation of ingested isoflavonoids. Total recovery of daidzein and genistein from urine was 35.8% and 17.6%, respectively. 4.4% of ingested daidzein, and 4% of genistein was recovered in feces. When O-DMA and equol were included, 54.7% of ingested daidzein was recovered. The plasma half life of daidzein was 6.0 hr, and that of genistein was 11.8 hr. Genistein seems to be more efficiently absorbed from the intestinal tract and to remain in circulation longer, so it may strongly interact with biomolecules in the body.

Various pharmacological effects of soy phytoestrogens are known. Genistein has been most widely studied. Various functions, such as inhibition of tyrosine kinase, hydroxysteroid dehydrogenase, 5 reductase, stimulation of steroids, hormone-binding-globulin production, inhibition of vascular proliferation, and angiogenesis have been identified (1-6). These pharmacological functions may prevent estrogen-related cancers. Because of their estrogenicity, they seem to be valuable also in other diseases caused by estrogen deficiency, such as osteoporosis and climacteric distress. On the other hand, excess intake of these substances may have a toxic effect, as described for -carotene (7). High consumption of soybean products in the Japanese diet results in high serum concentrations of daidzein and genistein, and high excretion of these isoflavonoids in the urine (8). We hypothesize that this is responsible for the low mortality from breast, endometrium, and ovarian cancer among the Japanese (9).

The aim of the present study is to summarize information regarding the pharmacokinetics of the absorption and excretion of isoflavonoids, and lignans, after a single oral dose of soybean powder (kinako) in humans.

Measurement of Isoflavonoids in Biological Specimens

Soybeans are well known to be rich in genistein and daidzein (10-12). In plants, almost all isoflavonoids are present in the form, glucoside ester. Fecal excretion of intact isoflavonoids was reported to be 0.4-0.7% in some subjects, and 6-8% in others. This difference is attributable to differences in the composition of intestinal bacterial flora (13). Isoflavonoids that possess a 5-OH group, such as genistein, are much more susceptible to C-ring cleavage by the intestinal bacteria of the rat (14). Certain strains of Clostridium in the human colon can cleave the C-ring and produce monophenolic compounds anaerobically (15).

It has been suggested that bacterial hydrolysis of the glycosidic link of the glycosidic forms of isoflavonoids must occur before absorption from the intestinal tract (15,16). O-DMA and equol, the main metabolites of daidzein, are probably produced in the colon (17,18)(Fig. 1). Hutchins et al. (19) compared urinary excretion following consumption of fermented and unfermented soy-foods, and found that the earlier absorption of phytoestrogens occurred by fermented soy.

Accurate measurements of plasma isoflavonoid levels are difficult, because of the small amounts in blood and complicated pretreatment leading to losses of the aglycone (20). Convenient HPLC UV detectors are not sensitive enough (usually 100-1,000 ng/ml). Gas chromatography-mass spectrometry (GC-MS) is the most sensitive method currently available, and allows detection in the order of 1 ng/ml, but it requires skillful technique (21). A recently developed electrochemical detector (HPLC-ECD) can detect levels on the order of 10-100 ng/ml (22). When measuring urinary isoflavonoids, vitamin C (final concentration is about 1%), and NaN_3 (final concentration 0.003 M) in storage bottles prevented disorganization of phytoestrogens and bacterial contamination.

In this study, plasma, urine, and feces samples were all analyzed by GC-MS methods in the Department of Clinical Chemistry, University of Helsinki (20,21). Eight different phytoestrogens, daidzein, genistein, O-DMA, equol, enterolactone, enterodiol, matairesinol, and secoisolacriresinol, were measured. Ingested kinako cakes contained an average 103 mole daidzein and 112 moles genistein after complete hydrolysis to aglycone.

The participants were requested to consume a soy product-free diet throughout the study period of 8 days. It was requested that tofu, natto (fermented soybeans), miso (fermented soybean paste) and shoyu be strictly avoided. It was also requested that lignan-rich foods, such as rye bread, whole grain bread, oatmeal, beer, bourbon whiskey, and soba noodles be avoided. Blood was collected 7 times, at 2-hr intervals, after ingestion of kinako, and each urine specimen was separately stored with a label indicating the time and amount. Feces were collected for 3 days, from Day 1 to Day 3. The experimental schedule is described in greater detail elsewhere (Watanabe et al., submitted).

Blood levels of isoflavonoids and lignans. Our previous analysis revealed very different levels of isoflavonoids in Japanese and Finns (8,23). In the present study, genistein and daidzein levels decreased to almost zero, after 3 days of avoidance of soybeans and soy products. Genistein peaked (2,440±651 nmol/L) 6 hours later after ingestion of kinako (Fig. 2). Plasma daidzein levels reached their level of 1,560±337 nmol/L at the same time. Genistein was predominant in 5 of the 7 subjects. Although some investigators have reported higher daidzein levels after soy milk consumption, daidzein dominance may depend on the kind of soybean food (24).

200

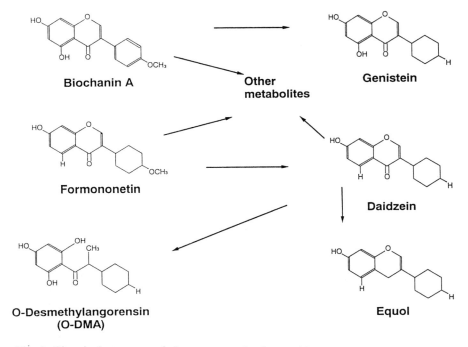

Biochanin A

Other metabolites

Genistein

Formononetin

Daidzein

O-Desmethylangorensin (O-DMA)

Equol

Fig. 1. Chemical structure of phytoestrogen isoflavonoids and metabolic pathway of daidzein and genistein.

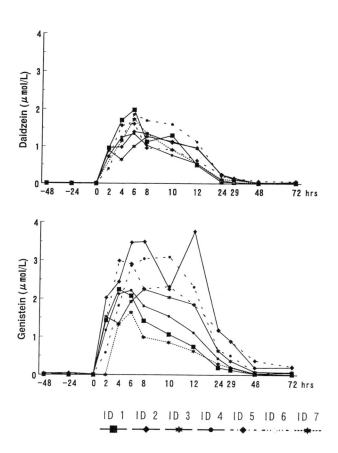

Fig. 2. Changes of plasma daidzein level (upper) and genistein level (lower). Each line indicates seven individuals. Subject 5 (ID 5) showed two peaks.

Plasma isoflavonoid levels decreased to their basal level 48 hours after ingestion of kinako. O-DMA and equol rose in one subject, whereas 3 of the 7 subjects showed low levels of these metabolites (low metabolizers). The regression line from the highest peak to the lowest level of genistein was given by the equation $y=-0.0965x+2.3578$ ($R^2=0.9798$), and the line for daidzein was given by the equation $y=-0.129x+1.551$ ($R^2=0.9906$). The half-life of plasma genistein calculated from this equation was 11.79 hr, and that of daidzein was 5.99 hr.

Fecal and Urinary Excretion of Isoflavonoids. The amounts of intact free daidzein and genistein in feces are very low. The conjugates, comprising less than 10% of the total, were not recovered. Fecal excretion of intact isoflavonoids was reported to be 0.4-0.7% in some subjects, and 6-8% in others, reflecting their plasma levels (13). Two of 7 women had high plasma levels, and excreted large amounts of isoflavone in their feces (13). No such tendency was observed in our study.

Fecal concentrations of isoflavonoids were often higher on Day 2 or 3. The marker color for kinako ingestion was recognized on Day 1 stools, suggesting that most fecal isoflavonoids came from conjugated forms excreted through the bile duct (2). Hydrolysis of conjugated phytoestrogens by intestinal bacteria may facilitate reabsorption.

Colon-liver circulation of isoflavonoids was more evident from their urinary excretion. Daidzein was the main component in urinary excretion. Its concentration started to increase shortly after the increase in plasma level, and reached 10,000 to 15,000 nmol/L 6 hours after kinako ingestion (Fig. 3). Almost all subjects characteristically showed two or three peaks during the 3 days, especially the low metabolizers (Fig. 4). Glycosidase and beta-glucuronidase activity in the feces was very low or absent in subjects who did not show a second peak in the urinary excretion, (Miura et al, in preparation).

In our experiment, subjects No. 1,3,4 showed increased O-DMA excretion after the peak of daidzein and genistein. Subject No. 6 excreted large amounts of equol after the peak of daidzein and genistein, and subject No. 7 excreted both equol and O-DMA. O-DMA is a major metabolite of daidzein (17,18). Equol is derived from various phytoestrogens (12). Equol did not show any significant association with other isoflavonoids by correlation analysis (Table 1). These differences must be attributable to differences in hepatic enzymes and/or intestinal bacterial flora.

Xu et al. (24) reported average 24-h urinary recovery of daidzein and genistein of 21 and 9%, respectively, at doses of 0.5-2.0 mg isoflavones/kg body weight after the single ingestion of soymilk. Average 48-hr urinary recoveries of ingested daidzein and genistein were 16 and 10%, respectively (13). Estimated bioavailability of isoflavones from soymilk varied from 13 to 35% depending upon the individual gut mircroflora (24). Our recovery rate was 54.7% for daidzein, including metabolites, and 20.1 % for genistein. Excess ingestion in Xu's study may have lowered absorption rate. Individual differences in total recovery ranged from 23.2 to 59.8% in our study (Table 2). The average level of isoflavonoids in plasma, urine, and feces are summarized in Table 3 (Table 3).

Conclusion and Future Problems

Interactions between isoflavonoids and biologically active molecules inside the body have not been investigated yet. Very little is known about absorption and transport through the intestinal wall. Most isoflavonoids are conjugated in the liver, and free phytoestrogens bind only to a low extent to SHBG in blood. Both free and sulfated forms are considered to be biologically active phytoestrogens. Recently, King et al.

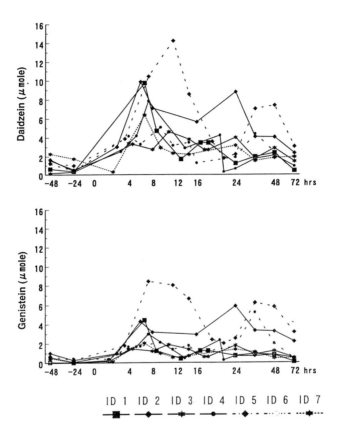

Fig. 3. Changes of urinary excretion of phytoestrogens. Daidzein (upper) overwhelms genistein (lower). Subject 5 (ID 5) excretes large amount of genistein, unlike others. Individual differences become larger than those of plasma isoflavonoids.

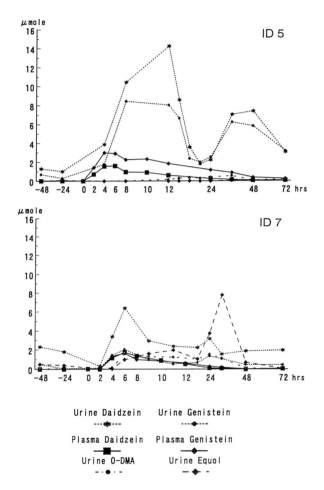

Fig. 4. Changes of plasma and urinary phytoestrogens in low metabolizer and high Metabolizer of daidzein. Subject 5 (upper) shows only a small elevation of O-DMA 12 hours after kinako ingestion (0 hr). Subject 7 (lower) is a high metabolizer of daidzein, and urinary equol excretion rapidly increased 16 hours after kinako ingestion. Two peaks in urinary excretion suggest the presence of enterohepatic circulation.

μmol increments) or ferulic acid (amounts ranging from 0 to 1 μmol, with 0.1 μmol increments) at 37 °C for 15 h.

In a separate experiment, samples containing blood plasma (516 μg of protein), 0.25 mmol of Trizma buffer (pH 7.4), 0.75 mmol of potassium chloride, 1 mmol of ferrous chloride, 0.5 mmol of hydrogen peroxide, and 10% of surfactant SDS were incubated with 2"-O-GIV (0, 0.05, 0.07, 0.1, 0.2, and 0.3 μmol) or probucol (0, 0.05, 0.07, 0.1, 0.2, and 0.3 μmol) at 37 °C for 15 h. In another experiment, samples containing blood plasma (860 μg of protein), 0.25 mmol of Trizma® buffer (pH 7.4), 0.75 mmol of potassium chloride, 1 mmol of ferrous chloride, and 0.5 mmol of hydrogen peroxide were incubated with 2"-O-GIV (amount ranging from 0 to 0.1 μmol with 0.01 μmol increment), vitamin C (amounts ranging from 0 to 0.1 μmol, with 0.02 μmol increments), or a mixture of 2"-O-GIV and vitamin C (total amounts ranging from 0 to 0.1 μmol, with 0.01 μmol increments), in which molar ratio of 2"-O-GIV and vitamin C was 1/1.

While the incubation continued, the mixtures were covered with parafilm. A 5-mL solution containing exactly the same materials except for the ferrous chloride and the hydrogen peroxide was incubated in the same manner as a control sample. Oxidation of the samples was stopped by adding 50 μL of 4% BHT ethanol solution. The incubation system was covered with aluminum foil to avoid any influence of light on the LDL and blood plasma peroxidation systems. The all experiments were replicated three times.

Analysis of Acetaldehyde and MDA. The amount of acetaldehyde was determined as a thiazolidine derivative according to the method reported previously (18). The amount of MDA was measured as a pyrazole derivative by the method developed previously (19). A Hewlett-Packard Model 5890 GC equipped with a nitrogen phosphorus detector (NPD) and a 30 m x 0.25 mm i.d. ($d_f = 1$ μm) DB-1 bonded-phase fused silica capillary column (J & W Scientific, Folsom, CA) was used for quantitative analysis of acetaldehyde and MDA. The detector and injector temperatures were 250 °C. The linear velocity of the helium carrier gas was 30 cm/sec with a split ratio of 21:1. The oven temperature was programmed from 60 to 180 °C at 4 °C/min and held for 10 min. GC peak areas were integrated with a Tsp SP 4400 series integrator. An HP Model 5890 series II GC interfaced to a HP 5971 mass spectrometer was used to confirm the identity of the thiazolidine derivative of acetaldehyde, 2-methylthiazolidine, and pyrazole derivative of MDA, 1-methylpyrazole, in the samples. The GC conditions were the same as for the GC with NPD. The mass spectra were obtained by electron impact ionization at 70 eV with an ion source temperature of 250 °C.

Results and Discussion

Oxidative damages caused by reactive oxygen species have been known to initiate and to promote many diseases, such as cancer (20), cardiovascular disease (21), and atherosclerosis (22). A strong relationship between atherosclerosis and amounts of lipid peroxidation products in the inside wall of arteries has been reported (23, 24). Recently, oxidative damage of LDL has received much attention as a process which implicates the development of human atherosclerosis (22). Therefore, LDL has been widely used to investigate the relationship between lipid peroxidation and atherosclerosis (25).

LDL is generally prepared from blood lipoprotein using a centrifuge. Figure 2 shows fractions of lipoprotein prepared according to their density (26). Figure 3 shows compositions of various lipoprotein fractions (26). LDL contains greatest

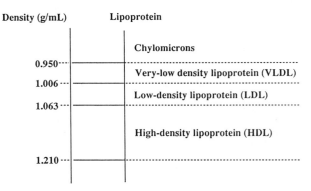

Figure 2. Fractions of lipoprotein.

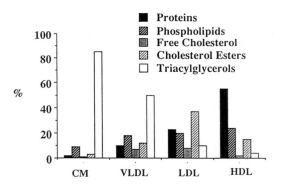

Figure 3. Compositions of various lipoprotein fractions.

Table 1: Total recovery of isoflavonoids from urine and feces (μmole)

Subject	Daidzein urine	feces	Genistein urine	feces	O-DMA urine	feces	Equol urine	feces	Total (%)**
#1	27.28	1.08	10.12	0.20	11.70	0.99	0.16	0.08	51.60 (24.0%)
#2	47.01	5.28	27.24	6.15	0.97	1.90	0.22	0.21	81.53 (37.9%)
#3	28.06	4.91	11.02	2.78	3.55	1.78	0.07	0.10	49.94 (23.2%)
#4	29.67	5.28	11.92	2.69	6.16	6.12	0.13	0.05	58.27 (27.1%)
#5	64.56	12.59	47.91	7.14	2.38	2.14	0.30	0.11	128.67 (59.8%)
#6	32.07	1.07	20.10	0.73	0.59	0.08	33.51	8.21	92.03 (42.8%)
#7	30.16	1.77	9.68	0.24	8.22	0.74	18.80	2.93	69.67 (32.4%)
mean	36.97	4.57	19.71	2.85	4.10	1.97	7.21	1.67	75.96
	35.8%	4.4%	17.6%	2.5%	4.0%*	1.9%*	7.0%*	1.6%*	35.3%

*these values are percent recovery of daidzein ingested. **% recovery of daidzein and genistein ingested.

Table 2. Correlation Coefficients between isoflavonoids(blood, urine, and feces)

	GEN_B	GEN_F	GEN_U	DAI_B	DAI_F	DAI_U
GEN_F	.6629	1.0000				
GEN_U	.6491	.8366**	1.0000			
DAI_B	-.1248	-.5250	-.0431	1.0000		
DAI_F	.3343	.8879***	.8337**	-.4378	1.0000	
DAI_U	.6036	.8814***	.9807***	-.1118	.8624**	1.0000
ODM_B	-.8074	-.6587	-.6131	.4270	-.4865	-.5253
ODM_F	-.1086	.2897	-.0326	-.6236	.3701	.0153
ODM_U	-.7532	-.5974	-.5761	.4153	-.4078	-.5073
EQ_B	.2642	-.3712	-.0200	.4427	-.4225	-.1801
EQ_F	.1067	-.4639	-.0912	.4865	-.4857	-.2231
EQ_U	-.0251	-.5237	-.1499	.5055	-.5181	-.2557

	ODM_B	ODM_F	ODM_U	EQ_B	EQ_F	EQ_U
GEN_B	-.8074**	-.1086	-.7532	.2642	.1067	-.0251
GEN_F	-.6587	.2897	-.5974	-.3712	-.4639	-.5237
GEN_U	-.6131	-.0326	-.5761	-.0200	-.0912	-.1499
DAI_B	.4270	-.6236	.4153	.4427	.4865	.5055
DAI_F	-.4865	.3701	-.4078	-.4225	-.4857	-.5181
DAI_U	-.5253	.0153	-.5073	-.1801	-.2231	-.2557
ODM_B	1.0000					
ODM_F	-.0512	1.0000				
ODM_U	.9506***	.0562	1.0000			
EQ_B	-.3488	-.4446	-.4218	1.0000		
EQ_F	-.1871	-.5214	-.3318	.9633***	1.0000	
EQ_U	-.0459	-.5420	-.2338	.8925***	.9807***	1.0000

GEN, genistein; DAI, daidzein, ODM, o-desmethylangolensin; EQ, equol; _B, blood; _F, feces; _U, urine. **, p<0.05; ***. p<0.001.

Table 3. Highest blood concentration, cumulative urinary and fecal excreted dose of Isoflavonoids

Variable	Mean	Std. Dev.	Minimum	Maximum
GEN_B	2.60	.71	1.64	3.76
GEN_U	19.71	14.03	9.68	47.91
GEN_F	2.85	2.82	.20	7.14
DAI_B	1.60	.27	1.27	1.98
DAI_U	36.97	13.90	27.28	64.56
DAI_F	4.57	4.04	1.07	12.59
ODM_B	.14	.14	.01	.34
ODM_U	4.80	4.10	.59	11.70
ODM_F	1.96	1.97	.08	6.12
EQ_B	.06	.14	.00	.37
EQ_U	7.60	13.37	.07	33.51
EQ_F	1.67	3.07	.05	8.21

Blood (μmole/L), Urine and Feces (μmole/3 days)

(25) reported changes in plasma, urine, and fecal concentrations after oral administration of soy extract and pure genistein to rats. Bacterial flora is present in the stomach of rats, so absorption starts early. Differences between bacterial flora in different species, however, must influence the biological effects.

Most phytoestrogens in plasma are present in their glucuronide form (8). These are excreted in bile, like human steroids, and deglucuronidation by intestinal bacteria would result in reabsorption from the colon. Circulation between colon-liver-colon may have a greater biological effect than previously considered.

The Japanese diet contains various soy products, including tofu, natto (fermented soybeans), kinako, miso, shoyu, etc. The insufficient estrogenic response in a recent dietary intervention study among postmenopausal women may have been due to the short period of intervention (4 weeks) (26).

The average intake of daidzein and genistein by Japanese is estimated to be 50 mg a day. Japanese also consume relatively large amounts of flavonoids (27). The antioxidant activities of flavonoids and isoflavonoids must take a part in oxidation-reduction reactions in the body (28-31). Continuous intake of these foods should keep plasma levels high, and contribute to preventing estrogen-related cancers, climacteric syndromes, and osteoporosis. Low colon cancer mortality is common in areas where consumption of soybean and soy products is high (9).

Acknowledgements

A part of this study was supported by a grant-in-aid for cancer research from the 10-Year Strategy for Cancer Control in the Ministry of Health and Welfare. The authors wish to thank Mr. T. Miura, Y. Arai, T. Takahashi, in the Tokyo University of Agriculture for their help in summarizing this work, and Drs. T. Sobue and M. Yamaguchi of the National Cancer Center Research Institute and National Institute of Nutrition and Health for facilitating this study.

References

1. Adlercreutz H, Fotsis T, Heikkinen R, Dwyer JT, Woods M, Goldin BR, Gotbach SL (1982) Excretion of the lignans enterolactone and enterodiol and of equol in omnivorous and vegetarian women and in women with breast cancer. Lancet 2: 1295-1299.
2. Adlercreutz H, Hockerstedt K, Bannwart C, Bloigu S, Hamalainen E, Fotsis, T, Ollus (1987). Effect of dietary components, including lignans and phytoestrogens, on enterohepatic circulation and liver metabolism of estrogens, and on sex hormone binding globulin (SHBG). J Steroid Biochem 27: 1135-1144.
3. Adlercreutz H, Goldin BR, Gorbach SL, Hockerstedt KAV, Watanabe S, Hamalainen EK, Markkanen MH, Makela TH, Wahala KT, Hase TA, Fotsis T. (1995) Soybean phytoestrogen intake and cancer risk. J Nutr 125: 757s-770s.
4. Fotsis T, Pepper M, Adlercreutz H, Flerischmann G, Hase T, Montesano R, Schweigerer L. (1993)
5. Genistein, a dietary-derived inhibitor of in vitro angiogenesis. Proc Natl Acad Sci USA 90: 2690-2694.
6. Messina MJ, Persky V, Setchell KDR, Barnes S. (1994) Soy intake and cancer risk: a review of the in vitro and in vivo data. Nutr Cancer 21: 113-131.
7. Setchell KDR, Borriello SP, Hulme P, Axelson M. (1984). Nonsteroidal estrogens of dietary origin: possible roles in hormone-dependent disease. Am J Clin Nutr 40:569-578.

8. van Genderen H. Adverse effects of naturally occurring nonnutritive substances. In; deVries J (ed), Food Safety and Toxicity. CRC Press, Boca Raton, New York, London, Tokyo, 1997, pp147-162.

9. Adlercreutz H, Markkanen H, Watanabe S. (1993). Plasma concentrations of phyto-oestrogens in Japanese men. Lancet 342: 1209-1210.

10. Watanabe S, Koessel S. (1993). Colon cancer: an approach from molecular epidemiology. J Epidemiol 3: 47-61.

11. Walter ED (1941) Genistin (an isoflavone glucoside) and its aglycone, genistein, from soybeans. J Am Chem Soc 63: 3273-3276.

12. Wang HJ, Murphy PA (1994) Isoflavone content in commercial soybean foods. J Agric Food Chem 42: 1666-1673.

13. Axelson M, Sjovall J, Gustafsson BE, Setchell KDR. (1984) Soya-a dietary source of the non-steroidal oestrogen equol in man and animals. J Endocrinol. 102: 49-56.

14. Xu X, Harris KS, Wang HJ, Murphy PA, Hendrich S. (1995) Bioavailability of soybean isoflavones depends upon gut microflora in women. J. Nutr. 125: 2307-2315.

15. Griffithzs, L.A., & Barrow, A. (1972). Metabolism of flavonoid compounds in germ free rats. Biochem. J. 130: 1161-1162.

16. Winter, J., Moor, J.H., Dowell, V.R., Jr. & Bokkenheuser, V.D. (1989) C-ring cleavage of flavonoids by human intestinal bacteria. App. Environ. Microbiol. 55: 1203-1208.

17. Brown JP (1988) Hydrolysis of glycosides and esters. In: Role of the Gut Flora in Toxicity and Cancer (Rowland IR, ed.), pp 109-144. Academic Press, San Diego, CA Eldridge A, Kwolek WF. (1983). Soybean isoflavones: effect of environment and variety on composition. J. Agric. Food Chem. 31: 394-396.

18. Yueh T-L, Chu H-Y. (1977) The metabolic fate of daidzein. Scientia. Sinica. 20:513-521.

19. Kelly GE, Nelson C, Waring MA, Joannou GE, Reeder AY. (1993) Metabolites of dietary (soya) isoflavonoids in human urine. Clin. Chim. Acta. 223: 9-22.

20. Hutchins AM, Slavin JL, Lampe JW. (1995) Urinary isoflavonoid phytoestrogen and lignan excretion after consumption of fermented and unfermented soy products. J. Am. Diet. Assoc. 95: 545-551.

21. Adlercreutz H, Fotsis T, Watanabe S, Lampe J, Wahala K, Makela T, Hase T. (1994) Determination of lignans and isoflavonoids in plasma by isotope dilution gas chromatography-mass spectrometry. Cancer Detect. Prev. 18:259-271.

22. Adlercreutz H, Fotsis T, Bannwart C, Wahala K, Brunow G, Hase T. (1991) Isotope dilution gas chromatographic-mass spectrometric method for the determination of lignans and isoflavonoids in human urine, including identification of genistein. Clin. Chim. Acta.199: 263-278.

23. Gamache PH, McCabe DR, Parvez H, Parvez S, Acworth IN. The measurement of markers of oxidative damage, anti-oxidants and related compounds using HPLC and coulometric array analysis. In; Progress in HPLC-HPCE Vol. 6, Acworth I.N., Naoi M, Parvez H, Parvez S (eds.),VSP, Utrecht, (1997) pp.99-126.

24. Adlercreutz H, Honjo H, Higashi A, Fotsis T, Hamalainen E, Hasegawa T, Okada H, (1991) Urinary excretion of lignans and isoflavonoid phytoestrogens in Japanese men and women consuming a traditional Japanese diet. Am. J. Clin. Nutr. 54: 1093-1100.

25. Xu X, Wang H-J, Murphy PA, Cook L, Hendrich S, (1994) Daidzein is a more bioavailable soymilk isoflavone than is genistein in adult women. J. Nutr. 124: 825-832.

26. King RA, Broadbent JL, Head RJ. (1996) Absorption and excretion of the soy isoflavone genistein in rats. J. Nutr.126: 176-182.
27. Baird DD, Umbach DM, Lansdell L, Hughes CL, Setchell KDR, Weinberg CR, Haney AF, Wilcox AJ, McLachlan JA (1995) Dietary intervention study to assess estrogenicity of dietary soy among postmenopausal women. J. Clin. Endocrinol. Metab. 80: 1685-1690.
28. Watanabe S, Arai Y, Kimira M. (1997). Uptake of flavonoids and isoflavonoids from food by Japanese. 16th Intnl. Congress of Nutrition, p118 (abs)
29. Pratt DE, Birac PM. (1979) Source of antioxidant activity of soybeans and soy products. J. Food Sci. 44: 1720-1722.
30. Gugler R, Leschik M, Dengler HJ. (1972) Disposition of quercetin in a man after single oral and intravenous doses. Angiologica 9: 162-174.
31. Hertog MGL, Feskens EJM, Hollman PCH, Katan MB, Kromhout D. (1993) Dietary antioxidant flavonoids and risk of coronary heart disease: the Zutphen elderly study. Lancet 342: 1007-1011.
32. Wei H, Bowne R, Cai O, Barnes S, Wang Y. (1995) Antioxidants and antipromotional effects of the soybean isoflavone genistein. Proc. Soc. Exp. Biol. Med. 208: 124-13

Chapter 20

Identification of Maillard Type Polymers with Antioxidative Activity

R. Tressll[1], G. Wondrak[1], E. Kersten[1], R. P. Kriiger[2], and D. Rewicki[3]

[1]Institut fur Biotechnologie, TU Berlin, Seestraβe 13, 13353 Berlin, Germany
[2]Institut fur Angewandte Chemie Adlershof e. V., Rudower Chaussee 5, 12484 Berlin, Germany
[3]Institut fur Organische Chemie, FU Berlin, Takustraβe 3, 14195 Berlin, Germany

In a series of ^{13}C-isotopic labelling experiments, Maillard reaction pathways of intact or fragmented sugars to N-alkylpyrroles and furan compounds were elucidated. The polymerizing activity of these compounds has been investigated. Complex mixtures of polymers and individual oligomers were characterized by FAB-/MALDI-TOF-MS and ^{1}H/^{13}C-NMR spectroscopy. The antioxidative activity of individual compounds was assessed by the Fe(III)-thiocyanate method and the DPPH-reduction assay.

The Maillard reaction, i.e. the nonenzymatic aminocarbonyl reaction between free or peptide-bound amines and reducing sugars, generally referred to as nonenzymatic browning, is one of the important reactions for food functionalization during thermal processing. Generation of colour, flavor and flavor enhancing compounds, as well as modification of texture, and enhancement of preservative qualities, are among the desired consequences of this reaction in food *(I)*. Moreover, several Maillard reaction endproducts of low molecular weight show remarkable bioactivity, e.g. reductones (2), metal chelators of the 3-hydroxypyridone type (3), dietary carcinogens of the imidazoquinoline type (4), and protein crosslinkers like pyrraline *(5)*.

Coloured melanoidins, the predominant Mailllayd reaction products (>95% p.w, depending on the model system) of the still unknown macromolecular structure, *(1,6,7)* contribute significantly to the functional properties of thermally processed foods: Melanoidins show remarkable antioxidative (reducing and radical scavenging) *(8),* desmutagenic (9), antibacterial *(10),* and protease inhibiting *(II)* effects. Low molecular coloured condensation products, isolated from Maillard model systems *(12,13),* do not necessarily represent melanoidin-like substructures.

Based on our previous work on reaction pathways of the Maillard reaction by [13]C- and [2]H-labelling experiments *(14,15)* we tried to throw light upon the structure of melanoidins(16). We expected pyrroles and furans to be intermediates, channelling low molecular weight Maillard products irreversibly into the macromolecular melanoidins. Therefore, in a series of model reactions we tested the identified title pyrroles and furans formed in the Maillard reaction, with respect to their polycondensation potential.

Browning of Maillard Reaction Systems

Naturally occurring sugars (D-glucose, D-ribose, 2-deoxy-D-ribose), as well as nucleic acids (DNA, RNA), show browning during long term incubation, with or without primary amines (Figure 1) *(16)*. Comparison of A_{420}, after a period of four weeks, demonstrates: (a) With amines (which, in general, induce much stronger browning), D-ribose is the most active sugar. (b) DNA shows stronger browning than RNA. Without amines, a comparable browning is observed with sugars, as well as with nucleic acids. With amines, the browning level is much lower in the latter case.

On one hand, our goal was to relate the superior browning activity of free or nucleic acids bound pentoses, to their generation by fragmentation from hexoses, and to their transformation to low molecular weight compounds as melanoidin precursors, on the other.

Fragmentation of Sugar Skeletons

Based on extensive ^{13}C-labelling experiments, detailed Maillard reaction scheme of hexoses and pentoses was elaborated (14,15). Thus, formation of products from intact, as well as from fragmented sugars, could be unambiguously attributed to distinct reaction pathways. A new pathway to pyrroles via a 3-deoxy-2.4-diketose intermediate (β-dicarbonyl route; Figure 2, intermediate **C**) was established. This route competes with the well known 3.4-dideoxyaldoketose route, via intermediate **B**, which is in accordance with the observed labelling patterns. Three different fragmentations *(α-dicarbonyl-, retroaldol-, and vinylogous retroaldol cleavage)* of the hexose intermediates **A-C** have clearly been revealed (Figure 2 and 3), resulting in formation of C_5 and C_4 Maillard compounds (derived from pentoses, 2-deoxypentoses and tetroses). Interestingly, the C_6-title compounds predominantly formed from the intact sugar skeleton (e.g. pyrraline from lysine), display only strong crosslinking reactivity, although their tendency to form melanoidins is low. Therefore, we focussed on C_5 and C_4 pyrrole and furan derivatives as suitable precursors for melanoidins.

Model Reactions to Melanoidin-like Polymers

Using synthetic N-methylpyrrole derivatives (and their ^{13}C-labelled analogs), which proved to be more suitable for analysis by CC-MS and MALDI-TOF-MS, we tested the endproducts shown in Figure 2 and 3 for the formation of polymers under mild conditions. MALDI-TOF-MS was chosen, because of its well documented analytical

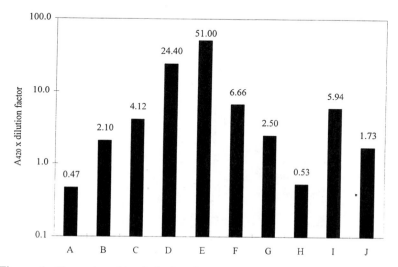

Figure 1. Browning (A_{420}) during long term incubation (40°C, pH 7, 0.5 phosphate buffer) of D-glucose (**A**), D-ribose (**B**), 2-deoxy-D-ribose (**C**), D-glucose + MAC (**D**), D-ribose + MAC (**E**), 2-deoxy-D-ribose + MAC (**F**), DNA (**G**), RNA (**H**), DNA + MAC (**I**), and RNA + MAC (**J**). MAC = methyl 6-aminocaproate.

212

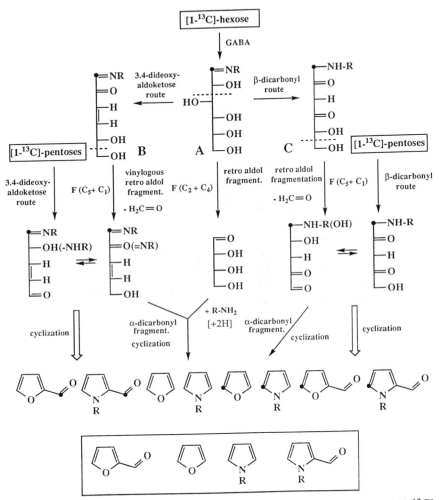

Figure 2. Formation of C5 and C4 compounds (pyrroles and furans) from [1-^{13}C]-D-hexoses by fragmentation and from [1-^{13}C]-D-pentoses (R = (CH$_2$)$_3$CO$_2$H) (*14, 15*).

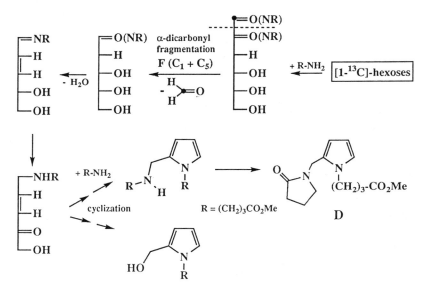

Figure 3. Transformation of [1-^{13}C]-D-hexoses into 2-deoxy-D-pentoses by α-dicarbonyl cleavage.

power for synthetic polymer research *(1 7,18)*. A systematic structural investigation by CC-MS, FAB-MS, MALDI-TOF-MS and ^{1}H-/^{13}C-NMR spectroscopy of single oligomers isolated by TLC, and of the polymers, revealed the formation of two distinct types of melanoidin-like polymers.

Formation and Structure of Type I Polymers. N-Methyl-hydroxymethyl-pyrrole originates from the Maillard reaction of methylamine, with 2-deoxy-D-ribose or (after *a-dicarbonyl* cleavage) from the reaction with hexoses, and could be identified as a trapped pyrrolidone derivative (**D**) from reaction of 4-aminobutyric acid with 2-deoxy-D-ribose (Figure 3) *(16)*. Moreover, a deformylated dimer P_2 and a trimer [1]P3 (Figure 4) were identified from the methylamine/2-deoxy-D-ribose system.

In a model reaction under mild conditions (20°C, H^+ catalysis), starting with synthetic N-methyl-2-hydroxymethylpyrrole (N-methyl-2-[^{13}C]-hydroxymethylpyrrole), repetitive electrophillic substitution and subsequent deformylation and dehydrogenation leads to the formation of linear methylene-bridged polypyrroles within minutes (Figure 4). This occurs in a biomimetic analogy to the porphyrinogen biosynthesis from porphobilinogen, a 2-aminomethyl substituted pyrrole. N-Substitution inhibits the cyclization of the tetramer, and linear species of up to 12 pyrrole unitsare generated (P_{12}-4 H). Starting from the hexamer (P_6), subsequent dehydrogenation with the formation of up to four conjugated pyrroles is *observed*. A brown to black color, and intense fluorescence (λex = 385 nm, λem = 493 nm, in *CHCl₃* develops while standing at room temperature. This is tentatively attributed to the chromophor of the methin-bridged polypyrroles, or to their corresponding radicals.

Compared to the outstanding polymerizing potential of N-alkyl-2-hydroxy-Methylpyrrole, the analogous furfurylalcohol is of minor reactivity. Even after 12 hours, the yield of oligomers (up to hexamers) is only moderate. The oligomers are either hydroxymethyl substituted furylmethanes or their ether-bridged isomers, undistinguishable by MALDI TOF MS. The polycondensation of furfuryl alcohol was the subject of several previous studies *(19,20)*.

Formation and Structure of Type II Polymers. N-substituted 2-formyl-pyrroles as well as N-substituted pyrroles, are common products in amino acid/hexose or pentose Maillard reactions. We studied N-methyl-2-formylpyrrole (N-methyl-2-[^{13}C]-formylpyrrole) (or furfural) in combination with N-methylpyrrole, as simple model systems under very mild conditions (20°C, H^+ catalysis, methanol), and observed extensive polycondensation within minutes (Figure 4). Polymeric species of up to about 25 methin-bridged pyrroles (or mixed species of up to 15 pyrroles and 14 furans) were detected. In the case of even numbered oligomers a subsequent dehydration/hydrogenation step is involved. A series of oligomers could be isolated and characterized by GC/MS, FAB/MS, ^{1}H-/^{13}C-NMR (Figure 5).

Due to the lower reactivity of the incorporated furan ring in electrophilic substitution, the polymer generated from N-methylpyrrole and furfural is obligatory linear. In contrast, the polycondensate from N-methyl-2-formylpyrrole and N-methylpyrrole may be branched. The intense red colour of the reaction mixture

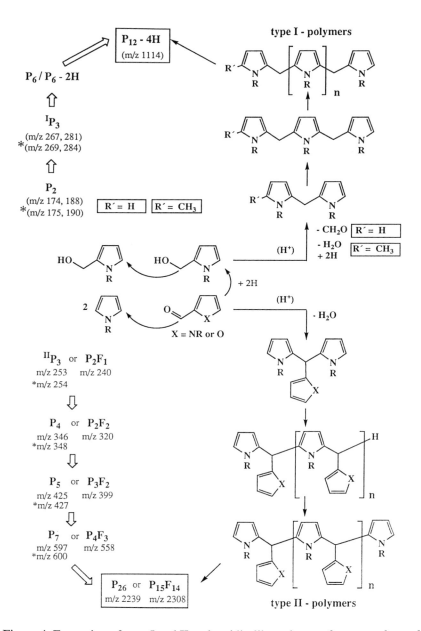

Figure 4. Formation of type I and II melanoidin-like polymers from pyrroles and furans (R = CH$_3$). **PnFm** = compound with n pyrrole and m furan moieties; * indicates m/z values of the labelled compounds.

(λmax = 515 nm, in CHCl$_3$) turned to brown/black upon exposure of the solid polymer to air. We attribute this colour to chromophors generated by dehydrogenation/oxidation processes. A subtype of type II polymers arises from the homopolycondensation of pure N-methyl-2-formylpyrrole, which is described in detail elsewhere.

With respect to the question whether the formation of type II polymers takes place in amino acid/reducing sugar sytems, or is inhibited by bulky amino acid side chain N-substituents, a mixture of N-(2-methoxy-carbonylethyl)pyrrole and N-(2-methoxy-carbonylethyl)-2-formylpyrrole (model compounds similar to those formed in the Maillard reaction of the Strecker inactive β-alanine) were tested. Interestingly, the polycondensation easily occurs, and a reaction product of type II polymer up to about 14 pyrrole units was detected.

Our model experiments demonstrate the formation of two distinct types of polymers from low molecular weight Maillard products and, therefore, presumably give a reasonable explanation for the dominant and irreversible transformation of the Maillard educts into macromolecular species. Of course, the structure of native melanoidins will differ from the described regular model polymers, but the competing polyreactions will possibly result in corresponding domains. Figure 6 summarizes the fragmentation pathways of different sugars leading to polymers of type I and type II. The important disaccharide specific fragmentation pathway (via vinylogous retroaldol cleavage) mentioned will be the subject of another paper.

Antioxidative Activity of Oligomers of Type I and II Polymers

To test the comparability of our model polymers and native melanoidins, we looked for coincidences in their functional properties. In particular, the structural similarity between the natural tetrapyrrole antioxidant bilirubin *(21)*, and the type I polymers, led us to investigate the antioxidative activity of selected oligomers. The tested oligomers, which were isolated chromatographically, are given in Figure 5. For comparison, standard antioxidants of different chemical constitution (phenolic, reductone, tetrapyrrole compounds) were chosen. Two types of assays for antioxidative activity were performed: (a) the DPPH-reduction assay according to *(22)* for assessment of the radical quenching activity and (b) the Fe(III)-thiocyanate assay according to *(23)*, which measures the suppression oflinoleic acid autoxidation.

Results. The DPPH-reduction assay (Figure 7) indicated a moderate radical quenching activity of the tested oligomers P_2, $^IP_3$ and $^{II}P_3$, whereas, their educts showed no activity . In the Fe(III)-thiocyanate assay (Figure 8), the relative antioxidant activity (in % = [l-Asample/Ablank] x 100) was assessed for two concentrations (1:10) of the test compound. The three standard antioxidants, n-octyl gallate, bilirubin and vitamin C, show strong activity at low and high concentrations. The educts show only very weak activity, The oligomers P_2, $^{II}p_3$, and P_2F_1 are strong antioxidants at both concentrations, whereas high concentrations of oligomers $^IP_3$ and P_3F_2 are needed. The oligomer P_5 is active in low concentrations, and shows pro-oxidant activity at higher concentrations .

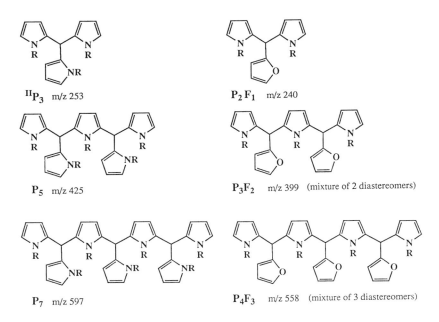

Figure 5. Isolated and characterized (FAB-MS, MALDI-TOF-MS and $^1H/^{13}C$-NMR) oligomers from N-methylpyrrole and N-methyl-2-formylpyrrole (furfural) polycondensations.

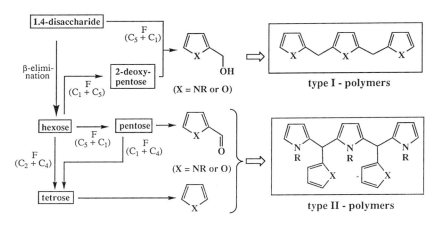

Figure 6. Maillard transformations of hexoses, pentoses and disaccharides into melanoidins (R = alkyl or amino acid residue).

218

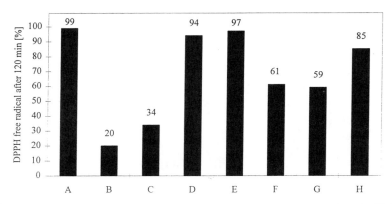

Figure 7. Antioxidative activity of samples **A** - **H** as assessed by the DPPH (2.2-di-[4-tert.octylphenyl]-1-picrylhydrazyl) reduction assay (*22*). **A** = blank, **B** = n-octyl gallate, **C** = vitamin E, **D** = N-methylpyrrole, **E** = N-methyl-2-formylpyrrole, **F** = dimer **P2**, **G** = trimer **^{I}P3**, **H** = trimer **^{II}P3**. Ordinate: DPPH concentration measured at λ = 516 nm (4 x 10^{-4} M = 100%). Sample concentration: 2 x 10^{-3} M.

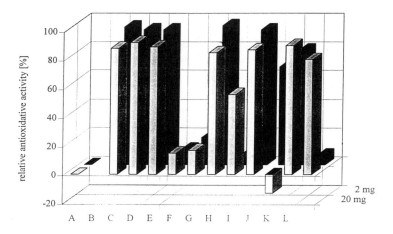

Figure 8. Antioxidative activity of samples **A** - **L** (2 mg / 20 mg) as assessed by the Fe(III)-thiocyanate assay (*23*). **A** = blank, **B** = n-octyl gallate, **C** =bilirubin, **D** = vitamin C, **E** = N-methylpyrrole, **F** = N-methyl-2-formylpyrrole, **G** = dimer **P2**, **H** = trimer **^{I}P3**, **I** = trimer **^{II}P3**, **J** = pentamer **P5**, **K** = trimer **P2F1**, **L** = pentamer **P3F2**. Ordinate: antioxidative activity in % = (1-A$_{sample}$/A$_{blank}$) x 100 measured at λ = 500 nm for two different amounts of test compounds (2 mg/ 20 mg).

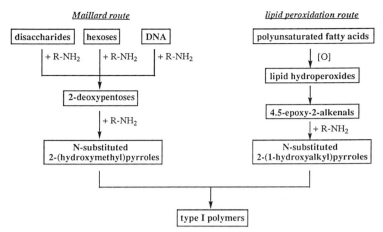

Figure 9. Convergence of the Maillard pathway and the lipid peroxidation/amine pathway (26) in the formation of type I melanoidin-like polymers.

Thus, the tested oligomers as representatives of the different types of model polymers display a pronounced antioxidative activity, which, in the case of the Fe(III)-thiocyanate assay, compares well with standard antioxidants. This antioxidative activity should be due to either the conjugative stabilisation of radicals by the adjacent pyrrole and furan rings formed after hydrogen abstraction from methylene- or methin-bridges, or their favored reaction with oxygen. Because of the poor solubility of the higher homologs, and the polymers in aqueous systems, and their laborious chromatographic preparation, the antioxidative activities of these species could not be determined.

Melanoidins from Aminocarbonyl Reactions

Recently, the linoleic acid autoxidation in the presence of amino compounds was studied (24-26). Interestingly, the key intermediate (E)-4.5-epoxy-(E)-2-heptenal generates N-substituted pyrroles by amino-carbonyl reaction with n-butylamine, or glycine methyl ester. These pyrroles undergo subsequent polycondensation with formation of oligomers and polymers, which are closely related to the type I polymers described above. The produced brown macromolecular pigments showed fluorescent characteristics similar to lipofuscins, the mammalian age pigments. Moreover, the products of a lysine/13-hydroperoxy-9(Z),11(E)-octadecadienoic acid reaction show antioxidative activity. From these results it was postulated that in food, as well as in mammalian tissue, fluorescent polypyrroles accumulate by amino-carbonyl reactions between free or protein-bound amino groups, and 4.5-epoxy-2-alkenals, from lipid peroxidation.

Our results demonstrate a simple and direct Maillard pathway to melanoidin-like polymers. Thus, two distinct reactions of extraordinary importance in food chemistry, the Maillard reaction and the lipid peroxidation in the presence of amines, seem to converge in the formation of dark coloured and fluorescent pigments of high molecular weight, with antioxidative properties (Figure 9).

Literature cited

1. Ledl, E; Schleicher, E. *Angew. Chem. Int. Ed. Engl.* **1990**, 29, 656.
2. Ninomiya, NI.; Matsuzaki, T; Shigematsu, H. *Biosci. Biotech.* **1992**, 56, 806.
3. Kontoghiorges, G.J.; Evans, R.W. *FEBS Letters* **1985**, 189, 141.
4. Sugimura, T. *Science* **1992**, 258, 603.
5. Nakayama, T Hayase, E; Kato, H. *Agric. Biol. Chem.* **1980**, 44, 1201.
6. Hayase, E; Kim, S.B.; Kato, H. *Agric. Biol. Chem.* **1986**, 50, 1951.
7. Benzing-Purdie, L.; Ripmeester, J.A.; Preston, C.M. *J. Agric. Food Chem.* **1983**,31, 913.
8. Hayase, E; Hirashima, S.; Okamoto, G.; Kato, H. *Agric. Biol. Chem.* **1989**, 53, 3383.
9. Lee, I.E.; Chuyen, N.V.; Hayase, E; Kato, H. *Biosci. Biotechnol. Biochem.* **1994**, 58, 18.
10. Einarson, H.; Eriksson, C. In *The MaillardReacn'on in Food Processing, Human Nutrition, and Physiology;* Finot, P.A.; Aeschbacher, H.U.; Hurrell, R.F; Liardon, R., Ed.; Birkhauser Verlag: Basel **1990**, 227-232.
11. Hirano, M.; Miura, M.; Gomyo, T. *Biosci. Biotechnol. Biochem.* **1994**, 58, 940.
12. A. Amoldi, E.A. Corain, L. Scaglioni and J. M. Ames *J. Agric. Food Chem.* **1997**, 45, 650.
13. E Ledl and Th. Severin Z. *Lebensm.-Unters. Forsch. 1978*, 167, 410.
14. R. Tressl, E. Kersten and D. Rewicki *J. Agric. Food Chem.* **1993**, 41, 2278.
15. R. Tressl, Ch. Nittka, E. Kersten and D. Rewicki *J. Agric. Food Chem.* **1995**, 43, 1163.
16. Wondrak, G.T.; Tressl, R. and Rewicki, D. *J. Agric. Food Chem.* **1997**, 45, 321.
17. Bahr, U.; Deppe, A.; Karas, M.; Hillenkamp, E; Giessmann, U. *Chem.* **1992**, 64, 2866.
18. Siuzdak, G. *Proc. Natl. Acad. Sci. USA* **1994**, 91, 11290.
19. Barr, J.B.; Wallon, S.B. *J. Appl. Polym. Sci.* **1971**, 15, 1079.
20. Wewerka, E.M.; Loughran, E.D.; Waiters, K.L. *J. Appl. Polym. Sci.* **1971**, 15, 1437.
21. Stocker, R.; Yamamoto, Y.; McDonagh, A. E; Glazer, A. N.; Ames, B. N. *Science* **1987**, 235, 1043.
22. Endo, Y.; Usuki, R.; Kaneda,T. *J. Am. Oil Chem. Sec.* **1985,** 62, 1387.
23. Inatani, R.; Nakatani, N.; Fuwa, H. *Agric.Biol. Chem.* **1983,** 47, 521.
24. Hidalgo, EJ.; Zamora, R. *J. Biol. Chem.* **1993**, 268, 16190.
25. Ahmad, I.; Alaiz, M.; Zamora, R.; Hidalgo, EJ. *J. Agric. Food. Chem.* **1996**, 44, 3946.
26. Zamora, R.; Hidalgo, EJ. *Biochim. Biophys. Acta* **1995**, 1258, 319. *Anal.*

Chapter 21

Tandem Mass Spectrometric Analysis of Regio-Isomerism of Triacylglycerols in Human Milk and Infant Formulae

H. Kallio, J.-P. Kurvinen, O. Sjövall, and A. Johansson

Department of Biochemistry and Food Chemistry, University of Turku, FIN-20014
Turku, Finland

Human milk lipids are generally recognized as an optimal source of fatty acids for both energy production and synthesis of functional lipids for infants. In addition to the special composition of fatty acids, combinations of fatty acids and their distribution within triacylglycerol molecules are also specific. Negative ion chemical ionization mass spectrometry was used to determine the molecular weight pattern and collision induced dissociation tandem mass spectrometry applied to identify the *sn*-2 *vs.* *sn*-1/3 positions of fatty acids in human milk and infant formula triacylglycerols. Significant differences were found between the products.

Mammal and marsupial dams exude milk for their neonates, as does a mother for her baby. The composition of the milk of various genera and species are analogous, though distinct differences do exist. Adaptation to different environments can be seen in milk composition. Whale calves swim in the ocean, kangaroo joeys are hidden in pouches whereas human children and their mothers try to survive in very variable climates and weather on dry land. The nutrient content of milk is also in balance with the species-specific speed of growth which is reflected by the proportions of proteins, lactose, lipids and minerals.

We may assume that the long evolution of mammals guarantees the milk of healthy, well-fed mothers to be optimal nourishment for their descendants. A man in modern society is no longer totally under the control of the natural selection of evolutive forces as in ancient times. Despite the rapidly changed and somewhat unnatural environment of infants, the composition of human milk remains the best food known for a child. Detailed evidence to the contrary has not been forthcoming.

Human milk is an excellent example of a functional food, even though the importance and detailed effects of its individual components are not yet fully understood. For the food industry, human milk is, however, the clear model for infant formulae.

Milk fat is a natural source of essential fatty acids and their precursors for cell membranes, eicosanoids and other functional lipids of infants. Fat is also needed for energy production and functions as an antimicrobial agent. In addition to the fat composition its digestion, absorption, resynthesis, transport, metabolism and placing in various tissues define the final physiological effect. The less developed lipase system of infants, high pH in the stomach, lipolytic enzymes in human milk and highly specific fatty acid positioning in the milk triacylglycerols make the fat metabolism of infants clearly distinct from that of adults.

The aim of the study was to compare the lipids of commercial infant formulae with mothers milk. Fatty acids, molecular weight distribution of triacylglycerols (TAG) and the regioisomeric structures of the major TAG were analysed by chromatographic and mass spectrometric methods.

Materials and Methods

Human Milk and Formulae. Milk was received from three Finnish mothers (28-34 years), about three months after delivery. The milk was pooled and frozen at -70 °C. Commercial brands of human milk formulae were either donations from producers or bought from retailers. The products were prepared according to the instructions on the packages. Lipids were isolated by a modified Folch extraction procedure with chloroform/methanol (2:1, v/v) (1) and triacylglycerols purified by a Florisil column, with 10 mL n-hexane/diethylether, (4:1, v/v). Results of the human milk sample, three infant formulae and one infant formula ingredient are shown and discussed as examples.

Fatty acid Analysis. Methyl esters of the TAG fatty acids were prepared by sodium methoxide transesterification (2). Gas chromatographic analysis was performed by a Perkin Elmer Auto Systems gas chromatograph (Perkin Elmer, Norwalk, Conn.), equipped with a split injector (split ratio 1:40, 230 °C) and a flame ionization detector (250 °C). An NB-351 column (25 m, 0.32 mm i.d., d_f 0.20 um, HNU-Nordion Instruments Ltd, Helsinki, Finland) was used with temperature programming from 50 °C to 240 °C at a rate of 4 °C/min and held at 240 °C for 10 min. The flow rate of the He carrier was 31 cm/s measured at 120 °C. Relative responses of various fatty acid methyl esters for the GC analysis were defined, using commercial mixtures GLC-60 and 68 D (Nu Chek Prep, Inc., Elysian, MN) as quantitative standards. Correction factors from methyl butanoate up to methyl docosahexaenoate varied between 1.15 and 1.00, where methyl hexadecanoate was used as the reference compound.

Mass Spectrometry of Triacylglycerols. The molecular weight pattern of TAG was determined by ammonia negative ion chemical ionization with a triple quadrupol tandem mass spectrometer (TSQ-700, Finnigan MAT, San Jose, CA) (3). Optimized parameters for relative abundances of [M - H]⁻ ions were used as published earlier (4). Each sample was analyzed four times and the averaged spectra displayed.

Tandem mass spectrometric analysis (TSQ-700) based on negative ion chemical ionization / collision induced dissociation was applied to the regioisomer analysis of TAG as described earlier (5,6). Triacylglycerols 16:0-18:1-16:0, *rac*-16:0-16:0-18:1,

18:1-16:0-18:1, *rac*-18:1-18:1-16:0 (Larodan Ab, Malmö, Sweden), *rac*-16:0-18:1-18:0, *rac*-16:0-18:1-18:2, *rac*-18:1-18:1-18:2 and *rac*-18:2-18:2-16:0 (Sigma Chemical Co, St Louis, MO) were used as calibration standards. Each of the standard TAG was analyzed four times and averaged results of both the correction factors of RCOO⁻ ions. The discrimination between the fatty acids in *sn*-2 and *sn*-1/3 of [M - H - RCOOH - 100]⁻ ions were determined for the default values in TAGS-100 programme based on the Simplex method (*6*). The correction factors were: stearic acid (18:0) 1.0; palmitic acid (16:O) 1.1; oleic acid (18:1n-9) 1.1 and linoleic acid (18:2n-6) 1.2. The averaged ratio of the abundances of [M - H - RCOOH$_{sn-2}$ - 100]⁻ / [M - H - RCOOH$_{sn-1/3}$ - 100]⁻ was 0.10 showing a typical discrimination, when using a rhenium wire loop introduction of the sample.

Results and Discussion

Fatty Acids. A gas chromatogram of fatty acid methyl esters of the pooled human milk sample, is shown as an example in Figure 1. Standard deviations of the ten major peaks typically varied between 1 and 4 %. Separation was sufficient, the only significant problem being the overlapping of fatty acids 22:6n-3 and 24:1n-9.

Fatty acid profiles of the five samples analyzed are shown in Table I. The pattern of human milk fatty acids corresponds well to the earlier investigations (*1,7,8*). The contents of fatty acids 4:0, 6:0 and 8:0 were higher than typically reported in the literature. This may be due to the special GC technique required for the proper analysis of short-chain fatty acids. However, Baldwin and Longenecker (1944) (*9*) have already correctly reported the existence of these compounds in human milk using fractional distillation techniques.

The wide compositional variation of the commercial baby milk formulae became clear in our investigations. The fatty acid profiles shown in Table I indicate the use of e.g. cow milk, vegetable oils and synthetic mixtures as ingredients. Formula I has a clear profile of cow milk fatty acids (*8*); Formula II is mainly based on cow milk. According to the fatty acid pattern, the formula ingredient oil could be related to the Formula III.

Molecular Weights of Triacylglycerols. Molecular weight patterns of triacylglycerols of the products are displayed in Figure 2. In addition to being fast, the MS method applied gives a more accurate molecular weight distribution than chromatographic methods. Comparison of the samples is easy and reliable. Figure 2a shows the typical pattern of human milk TAG, being in accordance with the results of e.g. Breckenridge et al. (*10*). Formula I (Figure 2b) is clearly based on bovine milk triacylglycerols, but Formula II (Fig 2c) may be composed of several milk TAG fractions, possibly supplemented with some vegetable oils. The molecular weight pattern of Formula III (Fig 2d) and the formula ingredient (Fig 2e) are practically identical. The only distinct difference being the high content of the ACN:DB species 52:2 of the ingredient. The molecular weight profiles of formulae II and III are similar to each other, but are clearly built up from different fat ingredients.

224

HUMAN MILK FATTY ACIDS

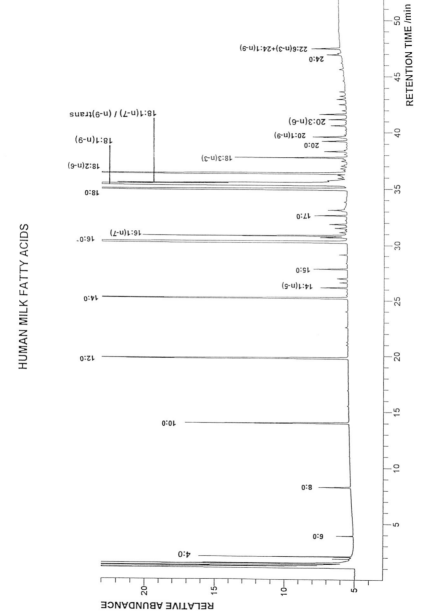

Figure 1. Gas chromatogram of methyl esters of triacylglycerol fatty acids of pooled human milk.

Table I. Relative proportions of TAG fatty acids of pooled human milk, formula ingredient fat and three commercial infant formulae.

Fatty acid (CN:DB)	Human Milk	Fat Ingredient	Formula I	Formula II	Formula III
4:0	0.2	-	9.2	4.9	-
6:0	0.2	0.3	4.3	1.9	0.6
8:0	0.4	2.9	1.9	0.9	10.1
10:0	2.0	1.8	2.9	1.5	5.2
12:0	6.6	13.0	2.8	1.9	15.4
14:0	6.9	5.3	8.2	4.8	5.3
14:1(n-5)	0.3	-	0.8	0.3	-
15:0	0.3	-	0.6	-	-
16:0	24.0	20.7	21.6	29.3	17.2
16:1(n-7)	2.3	0.1	0.9	0.6	0.1
17:0	0.3	-	0.3	-	-
18:0	10.3	2.3	7.7	5.5	4.0
18:1(n-9)	31.0	38.8	17.2	29.8	24.7
18:1(n-7)/(n-9)trans	1.7	0.8	1.0	1.4	0.6
18:2(n-6)	8.0	11.7	14.5	12.8	14.3
18:3(n-3)	1.2	1.2	2.3	1.4	1.3
20:0	0.3	0.2	0.2	0.3	0.2
20:1(n-9)	0.4	0.4	0.1	0.3	0.1
20:2(n-6)	0.2	-	-	-	-
20:3(n-6)	0.2	-	-	-	-
20:4(n-6)	0.3	-	-	-	-
20:3(n-3)	-	-	-	-	-
20:5(n-3)	0.1	-	-	-	-
22:0	-	0.1	0.1	0.1	0.2
22:1(n-9)	-	0.1	-	-	-
24:0	-	0.1	0.1	0.1	0.0
22:6(n-3)+24:1(n-9)	0.3	-	-	-	-
others	2.4	0.4	3.3	2.3	0.8
total (mol %)	100.0	100.0	100.0	100.0	100.0

226

a

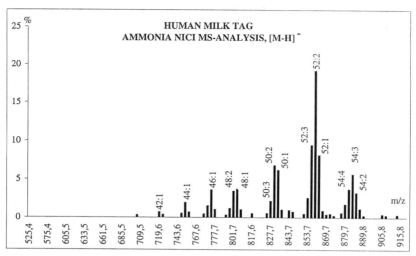

b

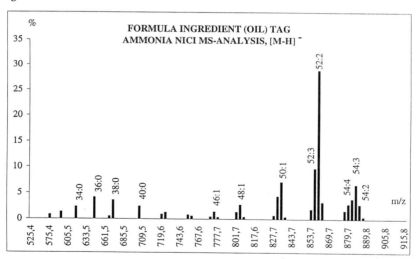

Figure 2. Molecular weight distribution of triacylglycerols of the five samples analyzed by ammonia negative ion chemical ionization mass spectrometry. 2a, pooled human milk; 2b, formula ingredient fat; 2c, Formula I; 2d, Formula II; 2e, Formula III.

c

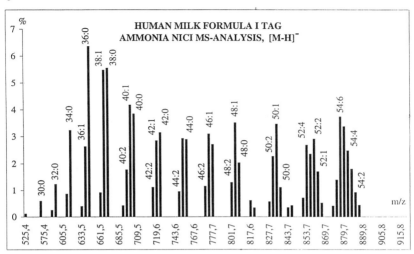

d

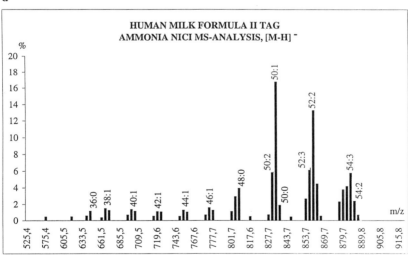

Figure 2. *Continued.*

Continued on next page.

e

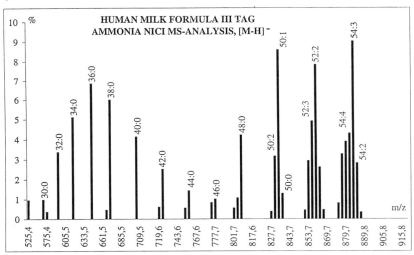

Figure 2. *Continued.*

Regioisomeric Structures of Triacylglycerols. The nutritionally often underestimated characteristics of fat in diets are the stereo- and regioisomeric forms of triacylglycerols. At each step in human fat digestion, absorption, resynthesis, transport, metabolism and placing in various tissues, positional discrimination between the isomers exists due to the different enzyme specificities. Positions of fatty acids within the molecules are a key factor in both catabolic and anabolic reactions. E.g. pancreatic lipase and plasma lipoprotein lipase hydrolyze preferentially glycerol esters of fatty acids in *sn*-1 and *sn*-3 positions instead of *sn*-2 position. The major biosynthetic pathway in the mucosal enterocytes is the 2-monoacyl-*sn*-glycerol pathway using the hydrolysis remnants, 2-monoacyl-*sn*-glycerols, as substrates.

The major ACN species in human milk was 52 (40 %) (Figure 2), consisting predominantly of three molecular weight species 856.7 (52:3), 858.7 (52:2) and 860.7 (52:1). These triacylglycerols of all five samples were analyzed by tandem mass spectrometry. Proportions of the regioisomers are listed in Table II.

All information concerning triacylglycerols in Table II was in good accordance with earlier known facts concerning human milk. The major TAG is *sn*-18:1-16:0-18:1 whilst *rac*-16:0-18:1-18:1 does not exist at all. In the TAG species 52:3 (16:0/18:2/18:1), palmitic acid is in *sn*-2 position, oleic acid mainly in the primary positions and linoleic acid exists in both primary and secondary positions. In the combination 16:0/18:1/18:0 (52:1) palmitic acid is again in the *sn*-2 position and stearic acid predominates the primary positions.

It is common knowledge that in human milk TAG palmitic acid is predominantly located in *sn*-2 position, palmitoleic and α-linolenic acids in *sn*-3 position, stearic acid in *sn*-1 position and oleic acid in both primary positions *sn*-1 and *sn*-3 (*10,11*). In

Table II. Percentages of regioisomers of triacylglycerols 52:1 (ACN:DB), 52:2
and 52:3 of the five samples analyzed by ammonia negative ion chemical ionization,
collision induced dissociation tandem mass spectrometry.

Triacylglycerol (ACN:DB)	Human Milk	Fat Ingredient	Formula I	Formula II	Formula III
52:1					
sn-18:1-16:0-18:0 + sn-18:0-16:0-18:1	98,2	89,0	65,3	34,0	39,4
sn-16:0-18:1-18:0 + sn-18:0-18:1-16:0	1,8	4,3	7,3	55,2	30,3
sn-16:0-18:0-18:1 + sn-18:1-18:0-16:0	-	6,8	27,4	10,9	30,3
52:2					
sn-18:1-16:0-18:1	92,2	81,8	49,0	20,2	30,4
sn-18:2-16:0-18:0 + sn-18:0-16:0-18:2	3,3	-	2,1	-	-
sn-16:0-18:2-18:0 + sn-18:0-18:2-16:0	4,5	-	11,3	-	5,9
sn-16:0-18:1-18:1 + sn-18:1-18:1-16:0	-	18,2	35,1	79,8	61,7
sn-18:2-18:0-16:0 + sn-16:0-18:0-18:2	-	-	2,5	-	2,0
52:3					
sn-18:2-16:0-18:1 + sn-18:1-16:0-18:2	76,8	80,8	16,1	24,6	33,4
sn-18:3-16:0-18:0 + sn-18:0-16:0-18:3	1,4	-	-	-	-
sn-18:1-16:1-18:1	10,0	0,5	1,5	-	-
sn-18:1-18:1-16:1 + sn-16:1-18:1-18:1	5,5	1,0	2,8	-	-
sn-16:0-18:2-18:1 + sn-18:1-18:2-16:0	3,7	6,6	62,6	51,3	35,9
sn-16:1-18:2-18:0 + sn-18:0-18:2-16:1	2,6	-	-	-	-
sn-16:0-18:1-18:2 + sn-18:2-18:1-16:0	-	11,1	17,1	24,1	30,7

230

addition, the positions of fatty acids in human milk TAG depend on the combination of the three fatty acids within the molecule (*5,6,12*). The tandem mass spectrometric analyses verify that stearic acid in human milk is very seldom located in *sn*-2 position whereas the incidence of myristic and palmitic acids in *sn*-2 position of individual molecular species varies between 20 and 90 %.

The corresponding structural characteristics (ACN 52) of the formula ingredient fat are close to the properties of human milk. Formula III which had fatty acids (Table I) and the molecular weight pattern of TAG (Fig. 2) close to the ingredient, showed completely different regioisomer composition and cannot be based on this ingredient fat. Formula I is a typical example of a bovine milk product, also in respect to the fatty acid distribution in ACN species 52:1, 52:2 and 52:3. Formula II also had characteristics of cow milk fat.

The tandem mass spectrometric method applied is a powerful tool when investigating the molecular structures of triacylglycerols for nutritional, technological and authenticity purposes.

Literature Cited

1. Jensen, R.G. *The Lipids of Human Milk*, CRC Press, Inc., Boca Raton, FL, 1989.
2. Christie, W.W. *J. Lipid Res.* **1982**, *23*, 1072-1055.
3. Kallio, H.; Currie, G. *Lipids* **1993**, *28*, 207-215.
4. Laakso, P.; Kallio, H. *Lipids* **1996**, *31*, 33-42.
5. Currie, G.J.; Kallio, H. *Lipids* **1993**, *28*, 217-222.
6. Kallio, H.; Rua, P. *J. Am. Oil Chem. Soc.* **1994**, *71*, 985-992.
7. Vuori, E.; Kinru, K.; Mäkinen, S.M.; Väyrynen, P.; Kara, R.; Kuitunen, P. *Acta Pediatr. Scand.* **1982**, *71*, 959-963.
8. Jensen, R.G. *Handbook of Milk Composition*, Academic Press, NY, 1995.
9. Baldwin, A.R.; Longenecker, H.E. *J. Biol. Chem.* **1944**, *154*, 255-265.
10. Breckenridge, W.C.; Marai, L.; Kuksis, A. *Can. J. Biochem.* **1969**, *47*, 761-769.
11. Breckenridge, W.C.; Kuksis, A. *J. Lipid Res.* **1967**, *8*, 473-478.
12. Kallio, H. *Proc. 17th Nordic Lipid Symp.* Imatra, Finland, June 13-15, 203 (1993).

Chapter 22

Effects of Various Phytochemicals on Colonic Cancer Biomarkers

B. A. Magnuson, J. H. Exon, E. H. South, and K. Hendrix

Department of Food Science and Toxicology, University of Idaho,
Moscow, ID 83844–2201

Epidemiological studies have demonstrated that populations
consuming high amounts of plant-based foods have a low incidence
of colonic cancer. We are currently utilizing a screening method
combining assays of biomarkers of colonic cancer and multiple
immune functions to identify the components of plants that may be
responsible for this protective effect. In several studies, rats were
fed one of the following compounds: chlorogenic acid (70 mg/kg),
quercetin (100 mg/kg), chlorophyllin (40 mg/kg), and curcumin (40
mg/kg). Colonic cell proliferation, the growth of preneoplastic
lesions (aberrant crypt foci) were evaluated as indicators of a
possible protective effect by the compounds. Although these
compounds have been reported as anticarcinogenic in at least one
cancer model, none of the compounds had a significant effect on
the colonic biomarkers tested. Therefore, we have not obtained
strong evidence to support the hypothesis that these compounds are
protective in the early stages of colonic cancer. The usefulness of
these biomarkers, and the influence of diet and study design are also
discussed.

The protective effect of fruits and vegetables against colonic cancer has been
demonstrated in many epidemiological studies (1, 2), resulting in considerable
research aimed at identifying the components responsible for this effect (3). We
evaluated the effects of dietary supplementation of a number of phytochemicals on
two biomarkers of colonic carcinogenesis, aberrant crypt foci and cell
proliferation, and on multiple immune functions in rats. The effects of these
compounds on the immune functions have been previously reported (4-6). This
report summarizes the effects of feeding chlorogenic acid (70 mg/kg), quercetin
(100 mg/kg), chlorophyllin (40 mg/kg), and curcumin (40 mg/kg) on colonic
biomarkers in rats.

Aberrant Crypt Foci. Aberrant crypt foci (ACF) were first observed by Bird *(7)* in whole mount colon preparations from carcinogen-treated mice and rats, and proposed as early preneoplastic lesions of colonic cancer. Considerable evidence has since been amassed to support this hypothesis. ACF have been observed in rodents treated with genotoxic *(8)* and non-genotoxic carcinogens *(9)*, and in humans with pre-existing or high risk of colonic cancer *(10, 11)*. Several investigators have reported a high frequency of K-ras mutations in ACF from humans *(12-15)* and rodents *(16)*. Compared to normal colonic crypts, ACF have increased cell proliferation rates *(13, 17, 18)*, increased carcinoembryonic antigen *(19)*, decreased hexosaminidase activity *(20)* and genomic instability *(21)*. Lack of expression of TGFα *(22, 23)* and decreased expression of TGFβ in ACF have also been reported *(23)*. These lesions are heterogeneous, with morphology ranging from near normal to severe dysplasia *(24-26)*. Changes in the growth of ACF have been extensively used as a means to predict tumor outcome and identify potential colon cancer preventive agents *(22, 27-30)*.

Chlorogenic acid. Chlorogenic acid is a plant phenolic found in a number of edible plants with higher concentrations in broccoli, potatoes and coffee beans *(31-33)*. Chlorogenic acid and some of its closely related plant phenolics such as ellagic acid, caffeic acid and ferulic acid have been shown to have chemopreventative activity in a number of animal tumor models *(34-36)*.

Quercetin. Quercetin is a flavonoid found in a variety of fruits, berries, vegetables, teas and red wines. Quercetin has been found to have mutagenic activity in several assay systems *(37)*. Quercetin has also been reported to have both promoting and inhibitory effects in various tumor models (for review see *37*). For example, dietary exposure to quercetin increased incidence of intestinal and bladder tumors *(38)*. Slight *(39)* to significant *(40)* increases in carcinomas of the large intestine of quercetin-fed rats have also been demonstrated. In contrast, dietary quercetin reduced mammary tumor development induced by DMBA and N-nitrosomethylurea in rats *(41)*. Azoxymethanol-induced colonic neoplasia in mice *(42)* and azoxymethane-induced ACF growth in rats*(43)* were reduced by dietary quercetin. Thus the effects of dietary QUE on cancer development are not predictive and may be related to the tissue or organ targeted by the carcinogen.

Chlorophyllin. Chlorophyllin (CHL) is a water soluble, food-grade derivative of the green plant pigment chlorophyll. This compound has also been reported to have both anticarcinogenic-like and cancer promoting activity. Chlorophyllin inhibited genotoxic activity of mutagens such as benzo[*a*]pyrene *(44-46)*, 3-methylcholanthrene *(46)*, 7,12-di-methylbenz(*a*)anthracene *(46)*, and aflatoxin B_1 *(47, 48)*. The incidence of hepatic cancer and DNA adduction was reduced in trout fed aflatoxin B_1 in combination with CHL *(47)*. Dietary supplementation of CHL delayed Zymbal's gland tumor development, reduced tumor burden in liver, Zymbal's gland and the small intestine, but shortened onset of skin tumors *(49)*. Mammary carcinogenesis induced by 2-amino-1-methyl-6-phenylimidazo[4,5-b]pyridine was inhibited in female rats fed 1% CHL in their diet, but the incidence of colon adenomas was increased *(50)*. CHL promoted dimethylhydrazine-induced colon tumors *(51)*, however inhibition by CHL of dimethylhydrazine-induced nuclear damage to rat colon epithelial cells had previously been reported *(52)*. Dietary CHL also inhibited formation of aberrant crypt foci and DNA adducts induced by 2-amino-1-methyl-6-phenylimadazo[4,5-b]pyridine in the rat colon *(49)*.

Curcumin. Curcumin is the major yellow pigment of tumeric and is found in curry and mustard preparations. Reports of the anticarcinogenic effects of curcumin include a reduction in azoxymethane-induced aberrant crypt foci *(30, 53)* and colonic tumors in rats *(54)*. A reduction in azoxymethane-induced prostaglandin formation was also reported in both liver and colon of the curcumin-fed rats *(53, 54)*. Huang and colleagues *(55)* reported oral administration of 2% curcumin reduced focal areas of dysplasia induced with azoxymethane.

The purpose of the studies summarized in this paper was to examine the effects of feeding these plant compounds on two biomarkers for colonic cancer, aberrant crypt foci growth and cell proliferation indices as a means of assessing their anticarcinogenic activity in the early stages of colonic carcinogenesis.

Methods

Animals. The animals were single housed in hanging wire cages in a temperature (20-22°C) and light controlled room (12hr on/off). Standard rodent chow (Teklad, Harland Labs) and deionized water were available *ad libitum* throughout the study. The rats were pretrained to consume graham crackers. A uniform (1.8-2.0 g) piece of cracker was then used as the substrate for addition of the various compound to the rat's daily food intake. The water soluble compounds (chlorophyllin, chlorogenic acid and quercetin) were mixed in 15% sucrose in water. Curcumin was solubilized in olive oil. The solutions were allowed to soak into the graham cracker before presentation and were fed 5 days/week (Mon. - Fri.) for the duration of the experiment. All animals consumed the cracker within 15 min of presentation with an average minimal crumb residue of <0.05 g. Rats were weighed twice a week throughout the experiments.

Chlorogenic acid studies. Four month old outbred male Sprague Dawley rats (Simonsen Laboratories Inc., Gilroy, CA) were used in the chlorogenic acid (CHA) studies. CHA (95% purity, Sigma Chemical Co., St. Louis, MO) was fed to the rats for 7 weeks at a daily dose of 70 mg/kg body weight. Two separate experiments were performed. In one, rats were divided into two groups of six rats each. One group received CHA as above and the other received only the VEH. The rats were terminated and immune function and colon proliferation and crypt height determined as described below. In the second study, rats were divided into two groups of 10 rats each. One group was treated with CHA as above and the other with VEH. Both groups of rats were also treated with azoxymethane (AOM) as described below.

Quercetin studies. Four month old outbred female Sprague Dawley rats (Simonsen Laboratories Inc.) were fed quercetin dihydrate (QUE)(Sigma Chemical Co.) at a dose of 100 mg/kg as described above for 7 weeks. Two groups of 8 rats received the diet supplement only (QUE or VEH) and 2 groups of 10 rats received AOM injections as described below and the dietary supplements (QUE-AOM and VEH-AOM).

Chlorophyllin studies. Outbred adult male Sprague Dawley rats (Simonsen Laboratories Inc.) were fed chlorophyllin (CHL)(Sigma Chemical Co.), as described above with final dose of 40 mg/kg for 7 weeks. Five animals per group received the diet supplementation only (CHL or VEH) and 2 groups of 10 rats received AOM injections as described below and the dietary supplements (CHL-AOM and VEH-AOM).

Curcumin studies. Adult outbred female Sprague Dawley rats (Simonsen Laboratories Inc.), seven per group, were used. Rats received either 40 mg/kg curcumin (Sigma Chemical Co.; >98% purity) or the olive oil vehicle only (VEH) weekdays for five weeks.

Aberrant crypt foci (ACF). Rats were injected sc with 20 mg/kg of azoxymethane (AOM) (Sigma Chemical Co.) in sterile water on day 2 and 9 of the feeding period. On the day of sacrifice for immune function assay, the entire colon was removed from each rat, flushed with saline, slit open and pinned flat on a balsa wood board and then submerged in 10% buffered formalin solution. The colon was stored in the formalin solution for at least 48 hr until preparation for evaluation of ACF. Visualization and quantification of the foci number and crypt multiplicity was accomplished as described by Bird (56). Briefly, the unsectioned colon was stained using 0.2% methylene blue (Sigma Chemical Co.) dissolved in phosphate buffered saline (pH 7.4) and whole mount segments examined under the light microscope. The colonic tissue was divided into 2 cm segments starting from the rectal end to determine distribution of ACF. The total number of foci per animal and the number of crypts per foci or crypt multiplicity were recorded. Foci composed of only one crypt were designated as AC1, foci composed of 2 crypts were designated as AC2, and so on.

The following safety precautions were taken by animal care personnel to reduce exposure to carcinogen metabolites excreted by the injected animals: gloves, mask, goggles and gown were worn and bedding was sprayed with soap/water solution to decrease dust particles. Bedding was disposed in boxes lined with plastic bags. All spills were cleaned thoroughly with soap and water.

Cell Proliferation. Proliferating cell nuclear antigen (PCNA) immunohistochemistry was used to detect changes in colonic cell proliferation as described (57) with modifications.

Immunohistochemistry. Segments of the colon of rats were taken from either both the ascending and descending colon or from the midpoint of the colon, and fixed in buffered formalin. Sections of embedded colon were deparafinized, rehydrated and treated with 1.5% hydrogen peroxide to quench endogenous peroxidase activity. Sections were blocked with 3% normal horse serum (Vector Laboratories Inc., Burlingame, CA) and then incubated with diluted mouse PCNA monoclonal antibody (Vector Laboratories Inc.) at a 1:200 dilution at room temperature for 1 hour. All incubations were carried out in a humidified chamber. Sections were incubated with secondary antibody, biotinylated antimouse IgG (rat adsorbed)(Vector Laboratories Inc.). Immunostaining was performed using the ABC method (Vector Laboratories Inc.) and 3,3'daminobenzidine tetrahydrochloride -hydrogen peroxide as the chromagen. A light Mayer hematoxylin counterstain was used.

   Scoring Colonic Crypts. Crypts were scored as described *(57)* with the following modifications. The number of cells containing either darkly staining nuclei (S phase) or lightly staining nuclei (G1 + G2 phase) *(58)* were recorded. The total number of cells in each crypt column (side) was defined as crypt height. The number of darkly staining cells divided by crypt height = labeling index. The number of all stained cells (dark + light) divided by the crypt height = proliferative index. Both indexes are expressed as a percent. Twenty values were obtained (10 crypts of 2 crypt columns each). The mean of these 20 values for each rat was then used in subsequent statistical analyses.

Statistical Analysis. Results are presented as means ± sem. Two group comparisons were analyzed for statistical significance using the Student T-test and one-way analysis of variance (ANOVA). Multiple group comparisons were performed using ANOVA followed by the Newman-Keuls *a posteriori* test for significant differences between individual means. $P < 0.05$ was considered significant.

Results

Body weight. In all studies, there was no significant effect of the test compounds on the body weights of animals (data not shown).

   Aberrant crypt foci (ACF). The total number of ACF and the percentage of aberrant crypts with different crypt multiplicity were not significantly different between AOM-treated chlorogenic acid (CHA)-fed and AOM-treated vehicle (VEH) controls. There was, however, a consistent trend toward elevated total ACF number and ACF with high multiplicity in the CHA-fed animals. Evaluation of the multiplicity and total number of ACF also showed no difference between chlorophyllin (CHL)-AOM and VEH-AOM rats (Table I). Although the quercetin (QUE)-AOM rats tended to have a higher percentage of ACF with high

236

multiplicity, this was only significant for ACF with a multiplicity of 6 (AC6). Total number of aberrant foci per colon (incidence) in QUE-AOM rats, was not significantly different from VEH-AOM rats (Table I).

Table I. Effect of dietary addition of chlorogenic acid (CHA), chlorophyllin (CHL) and quercetin (QUE) compared to vehicle (VEH) on percentage of foci with different crypt multiplicity and total number of aberrant crypt foci[1]

Treatment	AC1	AC2	AC3	AC4	AC5	AC6	AC>6	Total ACF
VEH-AOM	20.0± 5.5	43.8± 8.0	24.8± 4.9	9.0± 1.6	2.1± 0.6	0.3± 0.3	0.1± 0.0	111.4± 20.8
CHA-AOM	21.3± 2.9	40.6± 5.8	27.1± 3.0	8.5± 1.6	2.1± 0.9	0.4± 0.2	0.0± 0.1	127.5± 16.0
VEH-AOM	20.0± 3.3	43.8± 6.6	24.8± 3.4	9.0± 1.8	2.1± 1.0	0.3± 0.2	0.1± 0.1	111.4± 16.0
CHL-AOM	22.1± 5.2	39.0± 6.3	27.0± 4.8	8.7± 2.0	2.1± 1.1	1.1± 0.4	0.1± 0.1	110.5± 20.2
VEH-AOM	32.6± 5.6	41.8± 8.1	20.4± 4.7	4.4± 0.9	0.6± 0.2	0.0± 0.0[a]	0.0± 0.0	87.8± 15.6
QUE-AOM	30.7± 4.6	42.1± 5.6	21.0± 4.8	4.9± 1.3	0.9± 0.3	0.3± 0.2[b]	0.0± 0.0	101.2± 15.2

[1] All results are given as mean ± sem. N=10 for all groups except the VEH-AOM group for the CHL study, in which n=9. AC1 represents the percentage of total ACF which were composed of 1 crypt, AC2 represents the percentage of total ACF which were composed of 2 crypts, and so on.
[a,b] Results with differing letters are significantly different from corresponding VEH-AOM, $p<0.05$.
Data adapted from (4-6).

There was no difference in the distribution of ACF throughout the colon of CHL-AOM, CHA-AOM, and QUE-AOM rats as compared to their VEH-fed AOM controls (Table II).

Colonic cell proliferation indices. The colonic cell proliferation index and labeling index were consistently increased in CHA-treated rats compared to the control group but these effects were not statistically significant (Table III). There was no difference between CHA- and VEH-fed rats in average height of the colonic crypts (Table III).

Addition of QUE to the diet of rats did not alter indices of colonic epithelial cell proliferation. There were no significant differences in labeling index, proliferation index or height of the colonic crypts in QUE-fed rats compared to their VEH control (Table III).

Table II. Effect of dietary addition of chlorogenic acid (CHA), chlorophyllin (CHL) and quercetin (QUE) compared to vehicle (VEH) on the distribution of aberrant crypt foci (ACF) in the colons of azoxymethane-treated rats[1]

Cm from rectum	VEH-AOM [2]	CHA-AOM	VEH-AOM	CHL-AOM	VEH-AOM	QUE-AOM
2	5.7±2.2	7.8±2.6	5.7±2.2	5.1±1.5	6.0±1.5	7.8±2.5
4	12.3±3.6	16.5±4.4	6.7±1.9	6.3±1.4	12.5±3.0	13.7±2.6
6	8.3±1.9	12.2±3.6	8.3±1.9	11.0±3.2	8.9±2.4	14.8±3.3
8	11.4±3.1	14.3±4.7	11.4±3.1	19.5±4.5	12.5±3.8	18.3±3.6
10	17.2±5.8	14.7±3.8	17.2±5.8	18.3±7.0	14.0±2.8	17.4±4.4
12	14.2±3.4	14.5±3.1	14.2±3.4	13.1±4.5	12.9±2.8	13.3±2.9
14	18.4±4.1	25.2±4.8	18.4±4.1	21.7±5.5	15.2±4.8	12.2±3.4
16	22.7±5.0	20.8±4.2	22.7±5.0	11.9±4.1	5.4±2.0	2.8±0.9
18	3.8±1.1	7.5±2.8	3.8±1.1	2.3±0.9	0.4±0.3	0.9±0.7
20	2.6±1.2	1.8±0.9	2.6±1.2	1.0±0.5		
22	0.4±0.4	0.0±0.0	0.4±0.4	0.2±0.2		

[1] All results are given as mean ± sem.
[2] Number of ACF per 2 cm of colon.
Data adapted from (4-6).

Table III. Proliferation and labeling indices and average height of colonic crypts in rats fed chlorogenic acid (CHA), quercetin (QUE) or vehicle only (VEH) [1]

	VEH	CHA	VEH	QUE
Location	Mid-point	Mid-point	Mid-point	Mid-point
N	6	6	8	8
Proliferation Index[2]	4.9±0.5	7.4±1.2	8.5±1.2	7.3±0.6
Labeling Index	2.9±0.4	4.8±1.2	7.1±1.0	5.7±0.7
Crypt Height [3]	39.0±1.5	38.8±1.6	38.44±1.15	39.03±0.88

[1] All results are given as mean ± sem.
[2] Indices are expressed as a percent.
[3] Crypt height is the average number of cells from base to top of crypt.
Data adapted from (5, 6).

Chlorophyllin added to the diet of rats significantly increased the cell labeling index (p<0.03) and proliferating cell index (p<0.05) in epithelial cells of the ascending colon (Table IV). The ascending colon crypt heights were also greater (p<0.05) in chlorophyllin -fed rats when compared to VEH-fed rats. There was no significant difference between chlorophyllin - and VEH-fed rats in labeling index, proliferating cell index or crypt height in the descending colon.

Addition of curcumin to the diet of rats had no significant effect on any of the colonic cell proliferation indices in either the ascending or descending colon (Table IV).

Table IV. Proliferation and labeling indices and average height of colonic crypts in the descending (DC) and ascending (AC) colons of rats fed chlorophyllin (CHL), curcumin (CUM) or the corresponding vehicle (VEH) [1]

	VEH	CHL	VEH	CHL	VEH	CUR	VEH	CUR
Location	DC	DC	AC	AC	DC	DC	AC	AC
N	5	5	4	5	6	7	7	7
Proliferation Index [2]	9.5 ±1.0	12.2 ±3.6	8.1 [a] ±0.7	14.4 [b] ±1.6	9.1 ±1.5	9.1 ±2.6	7.8 ±0.6	9.3 ±0.8
Labeling Index [2]	5.4 ±1.0	7.2 ±2.0	4.6 [a] ±0.5	9.6 [b] ±1.6	5.8 ±1.1	5.7 ±2.1	7.8 ±0.6	9.3 ±0.8
Crypt Height [3]	32.3 ±0.6	31.8 ±1.1	28.2 [a] ±0.6	31.4 [b] ±1.3	30.2 ±0.7	30.1 ±0.8	30.5 ±0.8	30.3 ±0.8

[1] All results are given as mean ± sem.
[2] Indices are expressed as a percent.
[3] Crypt height is the average number of cells from base to top of crypt.
[a,b] Results with differing letters are significantly different, $p < 0.05$.
Chlorophyllin data adapted from *(4)*.

Discussion

We did not observe any significant effects of dietary addition of chlorogenic acid, quercetin or chlorophyllin on ACF growth in rats treated with azoxymethane in these studies. Similarly, the addition of chlorogenic acid, quercetin, and curcumin to the diet of rats had no significant effect on the indices of cell proliferation. The only significant effect observed in the colon was an increase in labeling and proliferation indices in the ascending, but not descending, colons of rats fed chlorophyllin. Therefore, we were unable to demonstrate any evidence of a trend towards reduced ACF growth or cell proliferation by these compounds. The trends observed were towards elevation of the number of ACF per cm in the colon, the total number of ACF per colon and the number of aberrant crypts per foci in the chlorogenic acid-treated group and quercetin-treated group as compared to their vehicle-treated controls. Although negative results (i.e. no significant differences) for one compound alone may not be remarkable, the consistent lack of effect of all of these compounds, which have previously been proposed as anticarcinogenic, provokes some thought and consideration. The importance of these negative results is that they may provide insight into the factors that affect the ability of dietary supplements to impact colon carcinogenesis.

There are several possible explanations for the consistent lack of inhibition of these four phytochemicals on the two colonic biomarkers employed in these studies. Firstly, the *compounds may not be protective* against colon cancer. For example, Pereira and coworkers *(29)* reported that neither the addition of chlorogenic acid (at 0.2 and 0.4 g/kg) or chlorophyllin (at 4.0 and 2.0 g/kg), to the diet altered the number or multiplicity of AOM-induced ACF in rats. Chlorophyllin was also reported to promote dimethylhydrazine-induced colonic tumors in rats *(51)*. These results are in contrast to reports of chemoprevention by chlorogenic acid and related compounds in other tumor models *(35, 36)*. Inhibition of formation of ACF and DNA adducts by the addition of 0.1% chlorophyllin to the drinking water of rats treated with 2-amino-1-methyl-6-phenylimidazo[4,5-b]pyridine has been reported *(59)*.

Studies with quercetin have also yielded conflicting results. The addition of the quercetin glycoside, rutin, at a level of 3.0 g/kg in the diet did not alter ACF growth *(29)*. Akagi et al. *(39)* found no effect of 2% dietary QUE on colon tumor development in rats. Also, feeding 1.68% dietary QUE for 10 weeks produced no significant effect on the number of AOM-induced ACF in rats, but ACF multiplicity was not reported *(40)*. In the same study, a significant increase in both incidence and multiplicity of adenocarcinomas was present in QUE-fed rats permitted to survive for 45 weeks after AOM treatment *(40)*. These results are in contrast to reports that both quercetin and rutin added to the diets inhibited AOM colonic tumor development in mice *(42)*. Recently, dietary addition of 2% quercetin was reported to inhibit AOM-induced ACF in rats *(43)*. The lack of a consistent reduction in colonic tumor development, plus reports of promotional activity *(39, 40)* and mutagenic activity *(37)* raise serious doubts regarding the suitability of quercetin as a potential anticancer treatment.

Curcumin added to the diet inhibited AOM-induced focal areas of dysplasia *(42)*, aberrant crypt foci *(30, 53)*, and colonic tumor development in rats *(54)* and mice *(60)*. Therefore, studies have consistently found the addition of curcumin to the diet to inhibit AOM-induced lesions.

Secondly, the *biomarkers used may not be sensitive or predictive* of colon cancer outcome. The evidence supporting the hypothesis that ACF are preneoplastic lesions of colon cancer was presented in the introduction. However, the reliability of either the number or the multiplicity of ACF as predictors of tumor outcome has been a subject of debate. In studies using cholic acid, the multiplicity but not the number of ACF was found to be predictive of tumor outcome *(28)*. Others have found neither multiplicity or number to be predictable *(61)*, however, possible reasons for the lack of correlation between ACF and tumor development in this study have been discussed in a recent review *(22)*. We only observed ACF development in these animals in this study for seven weeks. The effects of the compounds in this study on these preneoplastic lesions may have become more clearly defined following a longer latent period *(22, 28)*.

Cell proliferation indices have also been challenged as an indicator of colon cancer tumor outcome *(62)*. It should be noted that the cell proliferation data in this study were derived from groups that were not treated with azoxymethane. The results obtained from animals not treated with a carcinogen may not be applicable to carcinogen-treated animals. Curcumin added to the diet was reported to increase the number of thymidine-labeled cells in mice not treated with AOM, but had no effect on cell proliferation in AOM-treated animals (55). Increasing amounts of corn oil in the diet of rats had no effect on colonic cell proliferation in dimethylhydrazine-treated rats, but suppressed colonic cell proliferation in saline-treated rats (62).

Lastly, differences in the *experimental protocol* utilized may be responsible for the contrasting results. For example, the study by Guo et al. (49) which reported inhibition of ACF by chlorophyllin utilized 2-amino-1-methyl-6-phenylimidazo[4,5-b]pyridine as the carcinogen, whereas we used AOM in this study. Our findings, in combination with those of Pereira and coworkers (27), suggest that chlorophyllin may not be protective against AOM-induced lesions. One difference in our protocol compared to those used previously is presentation of the compound in a mid-day meal rather than incorporated into the diet for continuous exposure during the feeding period. The quercetin and curcumin doses in this study were lower than those used in continuous feeding studies that have reported decreased tumor incidence following carcinogen challenge. However, administration of the compounds in a single dose, as compared to ingestion as a percent of diet consumed over a day feeding period, may produce higher intraluminal concentrations at a single time point than those achieved by addition as a percent of total daily intake.

Supplementation of laboratory chow rather than the AIN-76 with various phytochemicals also represents a major difference in our studies compared to most previously reported studies. However, Matsukawa and coworkers also used laboratory chow as the base diet in their quercetin study (43). The use of laboratory chow makes comparison of experiments very difficult due to the variable nature of the micronutrient composition and phytochemicals from the plant ingredients of the laboratory chow. However, one can argue that unless the compounds are effective against a background of other phytochemicals in a variable diet, they are unlikely to be useful and effective additions to the human diet. Rats fed a commercial rat chow diet as compared to a purified diet containing cellulose exhibited altered lectin binding patterns (63), reduced neutral and sulfomucins (64), and altered intestinal pH (65). These observations indicate that the basal diet to which the phytochemicals or compound is added may modify the intestinal conditions and alter the effect of the compound on the biomarker being tested. This was observed with mice fed 0.2% cholic acid added to either the AIN-76 diet or laboratory chow (66). Mice fed cholic acid in the AIN-76 diet had significantly higher colonic cell proliferation indices than rats fed the AIN-76 diet alone (66). However, the reverse effect was observed when cholic acid was added to laboratory chow; mice fed cholic acid in the rat chow had significantly lower indices of cell proliferation (66).

In conclusion, we were unable to demonstrate any protective effect of these compounds on biomarkers of colon cancer. This may be due to one or a combination of the variations in the study protocol used. Clear elucidation of the factors which can alter the effects of dietary supplements on these biomarkers would further our understanding of the potential usefulness of these compounds as dietary supplements in the human diet.

Literature cited

1.   Trock, B.; Lanza, E.; Greenwald, P. *J. Natl. Cancer Inst.* **1990**, *82*, 650-661.
2.   Willett, W. C. *Environ. Health Persp. Suppl.* **1995**, *103*, 165-170.
3.   Kohlmeirer, L. In *Nutrition and biotechnology in heart disease and cancer*, eds. Longenecker, J. B., Kritchevsky, D., Drezne, M. C. Plenum Press, New York,1995, pp. 125-139.
4.   Exon, J. H.; Magnuson, B. A.; South, E. H.; Hendrix, K. *Environmental and Nutritional Interactions* **1997**, *1*, 83-95.
5.   Exon, J. H.; Magnuson, B. A.; South, E. H.; Hendrix, H. *J. Toxicol. Environ. Health* **1998**, *In press.*
6.   Exon, J. E.; Magnuson, B. A.; South, E. H.; Hendrix, K. *Immunopharm. Immunotoxicol.* **1998**, *In press.*
7.   Bird, R. P. *Cancer Lett.* **1987**, *37*, 147-151.
8.   McLellan, E. A.; Bird, R. P. *Cancer Res.* **1988**, *48*, 6183-6186.
9.   Whiteley, L.; Hudson, L., Jr.; Pretlow, T. *Toxicol Pathol* **1996**, *24*, 681-9.
10.   Pretlow, T. P.; Barrow, B. J.; Ashton, W. S.; O'Riordan, M. A.; Pretlow, T. G.; Jurcisek, J. A.; Stellato, T. A. *Cancer Res.* **1991**, *51*, 1564-1567.
11.   Roncucci, L.; Stamp, D.; Medline, A.; Cullen, J. B.; Bruce, W. R. *Hum. Pathol.* **1991**, *22*, 287-294.
12.   Pretlow, T. P.; Brasitus, T. A.; Fulton, N. C.; Cheyer, C.; Kaplan, E. L. *J. Natl. Cancer Inst.* **1993**, *85*, 2004-2007.
13.   Yamashita, N.; Minamoto, T.; Onda, M.; Esumi, H. *Jpn. J. Cancer Res.* **1994**, *85*, 692-698.
14.   Losi, L.; Roncucci, L.; di Gregorio, C.; de Leon, M. P.; Benhattar, J. *J. Pathol.* **1996**, *178*, 259-263.
15.   Shivapurkar, N.; Huang, L.; Ruggeri, B.; Swalsky, P. A.; Bakker, A.; Finkelstein, S.; Frost, A.; Silverberg, S. *Cancer Lett.* **1997**, *115*, 39-46.
16.   Vivona, A. A.; Shpitz, B.; Medline, A.; Bruce, W. R.; Hay, K.; Ward, M. A.; Stern, H. S.; Gallinger, S. *Carcinogenesis* **1993**, *14*, 1777-1781.
17.   Magnuson, B. A.; Shirtliff, N.; Bird, R. P. *Carcinogenesis* **1994**, *15*, 1459-1462.
18.   Pretlow, T. P.; Cheyer, C.; O'Riordan, M. A. *Int. J. Cancer* **1994**, *56*, 599-602.
19.   Pretlow, T. P.; Roukhadze, E. V.; O'Riordan, M. A.; Chan, J. C.; Amini, S. B.; Stellato, T. A. *Gastroenterology* **1994**, *107*, 1719-1725.
20.   Pretlow, T. P.; O'Riordan, M. A.; Kolman, M. F.; Jurcisek, J. A. *Am. J. Pathol.* **1990**, *136*, 13-16.
21.   Augenlicht, L.; Richards, C.; Corner, G.; Pretlow, T. *Oncogene* **1996**, *12*, 1767-72.

242

22. Bird, R. P. *Cancer Lett.* **1995**, *93*, 55-71.
23. Thorup, I. *Carcinogenesis* **1997**, *18*, 465-472.
24. McLellan, E. A.; Bird, R. P. *Cancer Res.* **1988**, *48*, 6187-6192.
25. McLellan, E. A.; Medline, A.; Bird, R. P. *Cancer Res.* **1991**, *51*, 5270-5274.
26. Siu, I.; Pretlow, T.; Amini, S.; Pretlow, T. *Am. J. Pathol.* **1997**, *150*, 1805-13.
27. Pereira, M. A.; Khoury, M. D. *Cancer Lett.* **1991**, *61*, 27-33.
28. Magnuson, B. A.; Carr, I.; Bird, R. P. *Cancer Res.* **1993**, *53*, 4499-4504.
29. Pereira, M. A.; Barnes, L. H.; Rassman, V. L.; Kelloff, G. V.; Steele, V. E. *Carcinogenesis* **1994**, *15*, 1049-1054.
30. Wargovich, M. J.; Chen, C. D.; Jimenez, A.; Steele, V. E.; Velasco, M.; Stephens, L. C.; Price, R.; Gray, K.; Kelloff, G. J. *Cancer Epidemiol. Biomark. Prev.* **1996**, *5*, 355-360.
31. Al-Saikhan, M. S.; Howard, L. R.; J.C. Miller, J. *J Food Sci.* **1995**, *60*, 341-347.
32. Huang, M.-T.; Ferraro, T. In *Phenolic Compounds in Food and Their Effects on Health*, Ed. Huang, M. T., Ho, C.T., Chang, Y.L. American Chemical Society, Washington, DC,1992, Vol. II, pp. 8-35.
33. Rodriguez de Sotillo, D.; Hadley, M.; Holm, E. T. *J Food Sci.* **1994**, *59*, 649-651.
34. Mori, H.; Tanaka, T.; Shima, H.; Kuniyasu, T.; Takahashi, M. *Cancer Lett.* **1986**, *30*, 49-54.
35. Tanaka, T.; Nishikawa, A.; Shima, H.; Sugie, S.; Shinoda, T.; Yoshimi, N.; Iwata, H.; Mori, H. *Basic Life Sci.* **1990**, *52*, 429-440.
36. Tanaka, T.; Kojima, T.; Kawamori, T.; Wang, A.; Suzui, M.; Okamoto, K.; Mori, H. *Carcinogenesis* **1993**, *14*, 1321-1325.
37. Formica, J. V.; Regelson, W. *Food Chem. Toxicol.* **1995**, *33*, 1061-1080.
38. Pamukcu, A. M.; Yalcimer, S.; Hatcher, J. F.; Bryan, G. T. *Cancer Res.* **1980**, *40*, 3468-3472.
39. Akagi, K.; Hirose, M.; Hoshiya, T.; Mizoguchi, Y.; Ito, N.; Shirai, T. *Cancer Lett.* **1995**, *94*, 113-121.
40. Pereira, M. A.; Grubbs, C. J.; Barnes, L. H.; Li, H.; Olson, G. R.; Eto, I.;Juliana, M.; Whitaker, L. M.; Kelloff, G. V.; Steele, V. E.; Lubet, R. A. *Carcinogenesis* **1996**, *17*, 1305-1311.
41. Verma, A. K.; Johnson, J. A.; Gould, M. N.; Tanner, M. A. *Cancer Res.* **1988**, *48*, 5754-5758.
42. Deschner, E. E.; Ruperto, J.; Wong, G.; Newmark, H. L. *Carcinogenesis* **1991**, *12*, 1193-1196.
43. Matsukawa, Y.; Nishino, H.; Okuyama, Y.; Matsui, T.; Matsumoto, T.; Matsumura, S.; Shimizu, Y.; Sowa, Y.; Sakai, T. *Oncology* **1997**, *54*, 118-121.
44. Elias, R.; Meo, M. D.; Vidal-Ollivier, E.; Laget, M.; Balansard, G.; Dumenil, G. *Mutagenesis* **1990**, *5*, 327-331.
45. Tachino, N.; Guo, D.; Dashwood, W. M.; Yamane, S.; Larsen, R.; Dashwood, R. *Mut. Res.* **1994**, *308*, 191-203.

46. Wu, Z. L.; Chen, J. K.; Ong, T.; Brockman, H. E.; Whong, W.-Z. *Teratogen. Carcinogen. Mutagen.* **1994**, *14*, 75-81.
47. Breinholt, V.; Hendricks, J.; Pereira, C.; Arbogast, D.; Bailey, G. *Cancer Res* **1995**, *55*, 57-62.
48. Breinholt, V.; Schimerlik, M.; Dashwood, R.; Bailey, G. *Chem. Res. Toxicol.* **1995**, *8*, 506-514.
49. Guo, D.; Horio, D. T.; Grove, J. S.; Dashwood, R. H. *Cancer Lett.* **1995**, *95*, 161-165.
50. Hasegawa, R.; Hirose, M.; Kato, T.; Hagiwara, A.; Boonyaphiphat, P.; Nagao, M.; Ito, N.; Shirai, T. *Carcinogenesis* **1995**, *16*, 2243-2246.
51. Nelson, R. L. *Anticancer Res.* **1992**, *12*, 737-740.
52. Robins, E. W.; Nelson, R. L. *Anticancer Res.* **1989**, *9*, 981-986.
53. Rao, C. V.; Simi, B.; Reddy, B. S. *Carcinogenesis* **1993**, *14*, 2219-2225.
54. Rao, C. V.; Rivenson, A.; Simi, B.; Reddy, B. S. *Cancer Res.* **1995**, *55*, 259-266.
55. Huang, M.-T.; Deschner, E. E.; Newmark, H. L.; Wang, Z.-Y.; Ferraro, T. A.; Conney, A. H. *Cancer Lett.* **1992**, *64*, 117-121.
56. Bird, R. P. *Cancer Lett.* **1995**, *88*, 201-209.
57. Yamada, K.; Yoshitake, K.; Sato, M.; Ahnen, D. J. *Gastroenterology* **1992**, *103*, 160-197.
58. Foley, J.; Ton, T.; Maronpot, R.; Butterworth, B.; Goldsworthy, T. L. *Environ. Health Persp. Suppl.* **1993**, *101*, 199-206.
59. Guo, D.; Schut, H. A. J.; Davis, C. D.; Snyderwine, E. G.; Bailey, G. S.; Dashwood, R. H. *Carcinogenesis* **1995**, *16*, 2931-2937.
60. Huang, M.-T.; Lou, Y.-R.; Ma, W.; Newmark, H. L.; Reuhl, K. R.; Conney, A. H. *Cancer Res.* **1994**, *54*, 5841-5847.
61. Hardman, W. E.; Cameron, I. L.; Heitman, D. W.; Contreras, E. *Cancer Res.* **1991**, *51*, 6388-6392.
62. Hardman, E. W.; Cameron, I. L. *Carcinogenesis* **1995**, *16*, 1425-1431.
63. Sharma, R.; Schumacher, U. *Lab. Invest.* **1995**, *73*, 558-64.
64. Sharma, R.; Schumacher, U. *Digest. Dis. Sci.* **1995**, *40*, 2532-9.
65. Ward, F. W.; Coates, M. E. *Lab. Anim.* **1987**, *21*, 216-222.
66. Robblee, N. M.; McLellan, E. A.; Bird, R. P. *Nutr. Cancer* **1989**, *12*, 301-310.

Chapter 23

Use of Cu(II)/Ascorbate System in the Screening for Hydroxyl Radical Scavenging Activity in Food Extracts via an ELISA for 8-Hydroxy Deoxyguanosine

Sachi Sri Kantha[1], Shun-ichi Wada[2], Masao Takeuchi[1], Shugo Watabe[2], and Hirotomo Ochi[1]

[1]Japan Institute for the Control of Aging, 723-1 Haruoka, Fukuroi City 437-01, Japan
[2]Laboratory of Aquatic Molecular Biology and Biotechnology, Graduate School of Agricultural Life Science, The University of Tokyo, Bunkyo ku, Tokyo 113, Japan

Conventional assays for measuring hydroxyl radical scavenging activity use either spin traps following the induction of the Fenton reaction, or measure thiobarbituric acid-reactive substances. The feasibility of using 8-hydroxy deoxyguanosine (8-OH dG), one of the sensitive and stable biomarkers for in vivo oxidative DNA damage, for measuring the hydroxyl radical scavenging activity, is described in this study. In this assay, 8-OH dG released from the hydroxylation of deoxyguanosine, by the Cu(II)/ascorbate system, is measured by competitive inhibition ELISA based on a specific monoclonal antibody to 8-OH dG.

Oxidative DNA damage has been described as a prominent cause of aging (1). One of the sensitive and stable biomarkers for in vivo oxidative damage is 8-OH dG (2). Kohen et al. (3) had described an in vitro system to quantitate 8-OH dG consisting of two steps. In the first step, deoxyguanosine was hydroxylated by a Cu(II)/ascorbic acid system to release 8-OH dG. This is based on the production of hydroxyl radicals from Cu(II), and ascorbate interactions (4). In the second step, the released 8-OH dG was measured by an HPLC-electrochemical detection (ECD) method. The sequences of chemical reactions, which occur in this interaction, are shown below.

$$\text{Ascorbate} + 2\text{Cu(II)} \text{-----------} \rightarrow \text{Dehydroascorbate} + 2\text{Cu(I)} + 2\text{H}^+ \quad (1)$$

$$\text{Ascorbate} + \text{O}_2 \xrightarrow{\text{Cu(II)}} \text{Dehydroascorbate} + \text{H}_2\text{O}_2 \quad (2)$$

$$H_2O_2 + Cu(I) \quad \text{---------} \rightarrow \quad Cu(II) + OH + °OH \qquad (3)$$

$$°OH$$
$$Deoxyguanosine \quad \text{---------} \rightarrow \quad \text{8-hydroxy deoxyguanosine} \qquad (4)$$

A competitive inhibition ELISA, based on a specific monoclonal antibody to 8-OH dG, which can be used as a measure to evaluate the hydroxyl radical scavenging activity in food extracts are described here. A preliminary report has appeared elsewhere (5).

Materials and Methods

Materials: The 8-OH dG standard was synthesized according to Kasai and Nishimura (6). Representative natural food extracts from beef, bonito, chicken meat, garlic and onion, with total solid content (as expressed by Brix value), ranging between 60 and 75, were obtained from Nikken Foods (Fukuroi City, Japan). The eel muscle extract was prepared from *Anguilla japonica* in the laboratory. Chemicals were purchased from Sigma Chemical Company, and used without further purification.

Optimal release of 8-OH dG from deoxyguanosine: Two optimization parameters examined were as follows: (a) The reactivity of various transition metal salts with ascorbic acid. For this experiment, the reaction mixture (1ml) contained 25 mM salt, 1 mM ascorbic acid, 10 mM deoxyguanosine and phosphate saline buffer. (b) The reactivity of Cu(II) with different organic acids. For this experiment, the reaction mixture (1ml) contained 25 mM $CuSO_4.5H_2O$, 1 mM organic acid, 10 mM deoxyguanosine and phosphate saline buffer.

All reactions were carried out in phosphate buffer/saline (8 mM $Na_2SO_4.12 H_2O$, 1.5 mM KH_2PO_4, containing 140 mM NaCl), pH 7.4, in a final volume of 1.0 ml. Table 1 shows the specific concentrations and volumes of each component of the reaction mixture.

Table I. Protocol for Liberation of 8-OH dG.

Component	Blank (μl)	Neg. control (μl)	Standard (μl)	Test compound (μl)
Phosphate/saline buffer	1000	800	700	600
Deoxyguanosine (10 mM)	---	---	100	100
$CuSO_4.5H_2O$ (100 mM)	---	100	100	100
Ascorbic acid (1 mM)	---	100	100	100
Antioxidant/food extract	---	---	---	100

As test compounds, representative antioxidants (other than bovine serum albumin) were used at 1 mM concentrations. The bovine serum albumin and solutions of natural food extracts were used at a 1% concentration. The reaction mixture was incubated at 37°C for 30 min., and the resulting product was subjected to ELISA for 8-OH dG. The degree of hydroxylation initiated by Cu(II)/ascorbate was calibrated by equating the release of 50 ng/ml of 8-OH dG, after 30 min. from the standard 10 mM deoxyguanosine as 100%.

ELISA for 8-OH dG: The functionality of the monoclonal IgG for 8-OH dG is discussed in Osawa et al. (7). The competitive inhibition ELISA was conducted on polystyrene, 96-well, flat bottom plates (Nunc-Immunoplate Maxisorb), coated by incubating with 100 ul of 8-OH dG, at 300 ug/ml overnight. The wells were washed with phosphate buffered saline (PBS). To prevent non-specific binding, 300 ul of 1% BSA-PBS was added to each well, and incubated for 2 hr. at room temperature.

Following a wash with PBS, quadruplicate samples (50 ul) of standard 8-OH dG, or test samples, were incubated with 50 ul of monoclonal IgG for 1 hr., at 37°C. This was followed by a wash with Tween 20-PBS, and addition of 100 ul horse radish peroxidase (HRP)-conjugated antimouse polyclonal IgG to each well. After an incubation for 1 hr at 37°C, the wells were washed again with Tween 20-PBS, and the bound HRP-conjugate was detected by reacting with substrate o-phenylenediamine plus H_2O_2 for 30 min. The reaction was stopped with 2N H_2SO_4 and the absorbance was measured at 492 nm using a computerized ELISA reader (MPR A4, Toyo Soda, Tokyo). The 8-OH dG standards used for the assay ranged between 0.64 and 2000 ng/ml. The concentration of 8-OH dG in the test samples was interpolated from the standard curve using log transformation.

Determination of carnosine: Carnosine content in food extracts was determined by the HPLC method described by Abe (8), using Zorbax 300-SCX (4.6 X 250 mm) column, with 90 mM KH_2PO_4 (pH 5.0) containing 10% methanol as an eluent.

Results and Discussion

Data presented in Fig. 1 and Table II validated the efficacy of the Cu(II)/ascorbic acid system studied by us in the release of 8-OH dG from deoxyguanosine. Fig. 1 shows that for the five transition metal salts examined, the efficacy in the release of 8-OH dG from deoxyguanosine decreased in the following order: Cu(II) > Fe(II) > Co(II) > Ni(II) >Mn(II). The amount of 8-OH dG released from deoxyguanosine in the Cu(II)/ascorbic acid system was almost five fold higher in comparison to the Fe(II)/ ascorbic acid system. The observed difference in the efficacy between Ni(II) and Mn(II) salts is marginal at best. When the reactivity of the Cu(II) ion was examined for comparison with ascorbic acid, with four other naturally occurring organic acids, the 8-OH dG release was not observed (Table II).

Table II. Reactivity of CuSO₄.5H₂O with various organic acids in the release of 8-OH dG from deoxyguanosine.

Ion/organic acid system	Released 8-OH dG (ng/ml)[a]
Cu(II)/ascorbic acid	71.4
Cu(II)/citric acid	0
Cu(II)/succinic acid	0
Cu(II)/maleic acid	0
Cu(II)adipic acid	0

[a]Each value is an average of duplicate determinations.

Somewhat comparable to our findings that formation of 8-OH dG from deoxyguanosine in the Cu(II)/ascorbic acid system is significantly higher than the Fe(II)/ascorbic acid system, Park and Floyd (9) had previously observed that compared to Fe(III), Cu(II) was more effective in mediating the 8-OH dG formation in calf thymus DNA by autooxidized liposomes under identical conditions. In addition, Chiou (4) had inferred that only Cu(I), Cu(II), Fe(II), and Fe(III) salts exhibit a significant DNA cleavage activity in the presence of ascorbate. He reported that the DNA-scission efficacy of different metal ions decreased in the following order:

Cu(I) and Cu(II) > Fe(II) > Fe(III) > Co(II) > Mn(II) and Ni(II)

The apparent resemblance of this DNA-scission efficacy profile of different metal ions with ascorbic acid, presented by Chiou (4), and our data on the 8-OH dG release efficacy profile of different metal ions with ascorbic acid, establishes the potency of the Cu(II)/ascorbic acid system as a useful tool for evaluating the hydroxyl radical scavenging activity.

Figure 2 shows the hydroxyl radical scavenging activities of representative antioxidants. Bovince serum albumin (1% solution), glutathione (1mM), and L-cysteine (1mM), showed 97-99% inhibition of 8-OH dG formed. On the same scale, the hydroxyl radical scavenging potential of carnosine (1mM) was equivalent to 89% inhibition potency of 8-OH dG production. Quantification of the antioxidant potential of carnosine, an endogenous dipeptide, was the primary objective for development of this method. Previously, Kohen et al.(3) reported that carnosine prevented the formation of 8-OH dG in the *in vitro* system, though they did not present a quantitative value. Subsequently, the hydroxyl radical scavenging activity of carnosine was demonstrated in quite a number of laboratories (10-13). We had

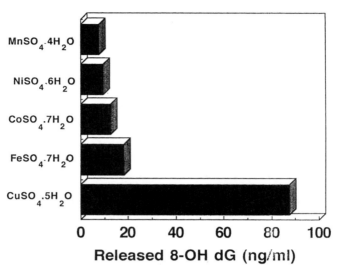

Fig.1

Reactivity of Five Transition Metal Ions with Ascorbic Acid in the release of 8-OH dG from deoxyguanosine. Each value is an average of duplicate determinations.

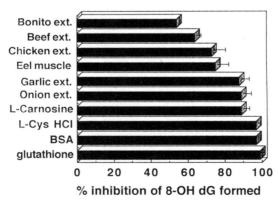

Fig.2

Screening of representative antioxidants and natural food extracts for hydroxyl radical scavenging activity. Each value is a mean of quadruplicate determinations (mean + SEM). Extract is abbreviated as ext.

recently shown that carnosine could significantly reduce the formation of 8-OH dG in rat embryonic fibroblasts, after four weeks of continuous culture(14). These previous findings, and the present data taken together, may suggest that the hydroxyl radical scavenging activity of carnosine could contribute significantly to its property of preventing the formation of 8-OH dG.

Among the six natural food extracts screened by the method described here, 1% solutions of onion and garlic extracts showed hydroxyl radical scavenging activity at 88-89% inhibition potency of 8-OH dG formation. Comparable values for hydroxyl radical scavenging activity were obtained for 1% solutions of eel muscle, chicken meat, defatted beef and bonito extracts, and were 75%, 73%, 64% and 54% respectively. The carnosine content in these four natural extracts was 103, 20, 180, and 25 umol/g respectively. It is not certain what percent of hydroxy radical scavenging potential is contributed by the carnosine present in these natural extracts. This is because carnosine is absent in the onion and garlic extracts. The 88-89% inhibition potency of 8-OH dG formation, detected in these two extracts, should be attributed to natural antioxidants other than carnosine.

The Fenton reaction, involving a Fe(II) salt and H_2O_2 has been studied in depth, relating to the potential of hydroxyl radicals in oxidizing a wide variety of organic acid substrates (15). We believe that the use of Cu(II)/Ascorbate system, via an ELISA for 8-OH dG, reported in this study, is a quick and sensitive screening method for hydroxyl radical scavenging activity in food extracts. We are currently calibrating on steps to extend the use of this ELISA method to test samples other than food extracts. Results of these studies will be published in a subsequent paper.

Literature Cited

1. Holmes, G.E, Bernstein, C. and Bernstein H. *Mutat. Res.,* **1992**, 275, 211-216.
2. Floyd, R.A, West, M., Eneff, K.L, Hogsett, W., Tingey, D.T. *Arch. Biochem. Biophys.,* **1988**, 262, 266-272.
3. Kohen, R., Yamamoto, Y., Cundy, K.C., Ames, B.N. *Proc. Natl. Acad. Sci. USA,* **1988**, 85, 3175-3179.
4. Chiou, S.H. *J. Biochem.(Tokyo),* **1983**, 94, 1259-1267.
5. Sri Kantha, S., Wada, S., Takeuchi, M., Watabe, S., Ochi, H. *Biotechnol. Techniques,* **1996**, 10, 933-936.
6. Kasai, H., Nishimura, S. *Nucl. Acid Res.,* **1984**, 12, 2137-2145.
7. Osawa, T., Yoshida, A., Kawakishi, S., Yamashita, K., Ochi, H. In *Oxidatives Stress and Aging,* R.G. Cutler, L.Packer, J.Bertram and A.Mori (eds), Basel, Switzerland, Birkhauser Verlag, **1995,** pp. 367-377.
8. Abe, H. *Bull. Jpn. Soc. Sci. Fish.,* **1981**, 47, 139.
9. Park, J.W., Floyd, R.A. *Free Rad. Biol. Med.,* **1992**, 12, 245-250.
10. Aruoma, O.I., Laughton, M.J., Halliwell, B. *Biochem. J.,* **1989**, 264, 863-869.
11. Rubtsov, A.M., Schara, M., Sentjure, M., Boldyrev, A.A. *Acta Pharm. Jugosl.,* **1991**, 41, 401-407.
12. Yoshikawa, T., Naito, Y., Tanigawa,T., Yoneta, T., Kondo, M. *Biochim. Biophys. Acta,* **1991**, 1115,15-22.

13. Chan, W.K.M., Decker, E.A., Lee, J.B., Butterfield, D.A. *J. Agric. Food Chem.,* **1994**, 42, 1407-1410.
14. Sri Kantha, S., Wada, S., Tanaka, H., Takeuchi, M., Watabe, S., Ochi, H. *Biochem. Biophys. Res. Comm.,* **1996**, 223, 278-282.
15. Walling, C. *Acc. Chem. Res.,* **1975**, 8, 125-131.

Chapter 24

Effect of Oxygen Depletion on the Food Digestion

R. Yamaji, M. Sakamoto, Y. Ohnishi, H. Inui, K. Miyatake, and Y. Nakano

Department of Applied Biological Chemistry, Osaka Prefecture University, Sakai, Osaka 593, Japan

Hypoxemia causes anorexia, and consequently results in weight loss. Under chronic hypoxic conditions, weight loss occurred even when food intake was adequate, indicating that hypoxia, but not anorexia, resulted in weight loss. Furthermore, when conscious rats were acutely exposed to normobaric hypoxia (10.5% or 7.6% O_2), hypoxia inhibited the gastric emptying and the gastric acid secretion and increased the plasma gastrin concentration. Thus, acute hypoxemia lowers gastric function. However, the hypoxia-suppressed gastric acid secretion was gradually recovered during chronic exposure to hypoxia.

Hypoxemia, which results in a reduction of arterial oxygen pressure in the blood, and a lack of oxygen in tissues, is a biological state that mammals, including humans, may encounter acutely, or chronically. It may come by way of cardiopulmonary diseases, pernicious anemia, and residence at high altitudes. Hypoxemia resulting from various clinical situations, and trekking or expeditions to high altitudes, is associated with weight loss or failure to gain weight. Several possible explanations for this phenomenon are; 1) a decreased food intake due to anorexia (1-4); 2) an increased energy requirement due to an increased basal metabolic rate (5-8); 3) a loss of body water through diuresis (1, 9), or through increased respiratory and insensible losses (10); and 4) a decreased absorption and digestion (1, 4, 10-16). The amount and rate of weight loss seems to depend on the length of hypoxia (short- or long-term), and the degree of hypoxia (severe or mild), suggesting that weight loss via hypoxia varies with the experimental procedures. Although several investigators have studied absorption tests in people with hypoxemia resulting from lung and heart diseases, pernicious anemia and residence at high altitude, contradictory results for both absorption and digestion have been reported. We report in this paper, that chronic hypoxia results in weight loss, and that acute hypoxia inhibits a gastric emptying and a gastric acid secretion when conscious rats are exposed to 10.5% or 7.6% O_2 hypoxia.

252

Weight Loss and Malabsorption

Hypoxemia causes anorexia, and consequently results in weight loss (1-4). To study the effect of hypoxia on dietary intake, rats were exposed to 7.6% O_2 hypoxia. Before the experiment began, five-week-old male Wistar rats, weighing 110-120 g, were housed individually in stainless steel cages. They were adapted to a meal-feeding schedule, which allowed free access to a nonpurified rat diet (CE2: Clea Japan, Tokyo) for 4 h (from 1100 to 1500 h) every day, for 2 weeks. On the day of the experiment, rats were transferred to 7.6% O_2 hypoxic chamber immediately, before 4 h of consuming diet. The amount of diet, which the hypoxic rats consumed for 4 h, was weighed. As shown in Figure 1, food intake decreased after hypoxic exposure and then was gradually recovered to near the normoxic level during 5 d of hypoxia.

Next, the effect of hypoxia on body weight was studied. Body weight in the hypoxic rats was compared to that in the normoxic rats, which were pair-fed. As shown in Figure 2, weight loss due to the decreased food intake was observed after 2 days of hypoxia, whereas there were no significant differences in weight between the hypoxic rats and the normoxic rats with pair-feeding. In contrast, when rats were chronically exposed to hypoxia (more than 3 wk), weight gain in the hypoxic rats was less than that in the normoxic rats with pair-feeding. These results indicate that chronic hypoxia causes weight loss even when food intake is adequate. Thus, chronic hypoxia results in weight loss by factors except for food intake. Absorptive capacity may be associated with weight loss in hypoxia. Therefore, a xylose absorption test has been performed. Milledge (17) found that there was a definite correlation between xylose absorption and arterial oxygen saturation in patients with hypoxemia resulting from chronic respiratory disease, or congenital cyanotic heart disease. Consequently, he predicted that similar changes in xylose absorption occurred in sojourners who experienced decreased arterial oxygen pressure at high altitudes. Boyer and Blume (1) measured urinary xylose excretion during exposure to altitude of 6,300 m in the 1981 American Medical Research Expedition to Everest and found that hypobaric hypoxia decreased the urinary xylose excretion, compared to the sea level. At this altitude, fat absorption also decreases. However, no fat malabsorption at altitudes >4,700 m and no protein malabsorption at altitudes <5,000 m are observed (15, 16). Furthermore, the urinary xylose excretion in sojourners for 4, 10 and 20 days at an altitude of 3,500 m is the same as that in residents at sea level (4). Thus, contradictory results have been showed in studies of hypoxemia. Possible causes for contradictory results may be a different degree of hypoxia (severe or mild) and a different length of hypoxic exposure (short-term or long-term).

Urinary Nitrogen Excretion. Although no protein malabsorption is found at high altitudes, the urinary ^{15}N excretion seems to increase at a high altitude (5,000 m) when [^{15}N]glycine is intravenously administered (15). The increased rate of production and/or excretion of creatinine, which is one of metabolites of glycine, is proposed as one possible explanation for this phenomenon. However, the 48-hours cumulative urinary urea excretion is greater in the hypoxic group than in the normoxic group when rats are exposed to normobaric hypoxia (10.5% O_2 or 7.6% O_2) immediately after feeding diet (18). The increased urinary ^{15}N at a high altitude may be attributed to urea rather than creatinine, because glycine is also a precursor of urea, and most of the nitrogen from protein and/or amino acid catabolism is normally lost as urea through urinary excretion. Thus, hypoxia may accelerate the turnover rate of intrinsic proteins and/or amino acids,

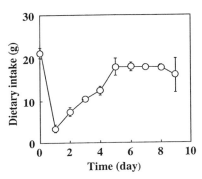

Figure 1.   Effect of hypoxia on dietary intake .   Values are expressed as means _± SEM of 15 rats.

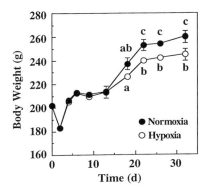

Figure 2. Effect of hypoxia on body weight. Rats were kept according to the methods as described in Figure 1. The pair-fed rats were kept in the normoxic chamber as a control group.   Values are expressed as means ± SEM of 15 rats/group. Values with different letters superscripts are significantly different ($P < 0.05$).

or cause greater absorption of peptides and/or amino acids in the intestine, whereas the latter explanation is in conflict with no protein malabsorption at high altitudes.

Gastric Emptying and Gastric Acid Secretion.   The principal function of the stomach is to act as a reservoir of ingested foodstuffs which are slowly emptied into the small intestine, in the optimum size and in proportion to digestive juices, to enhance nutrient absorption.   No changes in the digestibilities of carbohydrates and fats are found at high altitudes (2, 4, 14, 16).   In contrast, it has been reported that gastric emptying and gastric acid secretion are inhibited in patients with hypoxemia resulting from pernicious anemia (19), and that gastric acid production is lower in natives at 3,500 m, sojourners for 22 days at 3,500 m, and residents for 1 year at 3,500 m, than in residents at the sea level (4).

When rats are exposed to normobaric hypoxia at 10.5% O_2 or 7.6% O_2, gastric emptying is inhibited (18).   Furthermore, to study the effect of hypoxia on gastric acid secretion, conscious rats were subjected to pylorus ligation and were acutely exposed to 7.6% O_2 or 10.5% O_2.   Rats had free access to water up to 1 h before operation. While under light ether anesthesia, abdomens of unfed rats were opened through a ventral median celiotomy, and the pylorus was ligated.   Rats regaining the righting reflex 5-10 min after the operation were transferred to the hypoxic chamber (10.5% or 7.6% O2), or the normoxic chamber.   Three hours later, gastric acid output was determined by back-titration to pH 7.0 with 0.1 N NaOH.

As shown in Figure 3, the gastric acid secretion was lower in the two hypoxic groups than in the normoxic group and consequently the gastric pH was greater in the hypoxic rats than in the normoxic rats.   However, when rats were chronically exposed to 7.6% O_2 hypoxia for 7 or 9 days and further exposed to 7.6% O_2 for 3 hr after pylorus ligation to measure the gastric acid secretion and the gastric pH, the gastric acid and the gastric pH increased and decreased, respectively, compared to those in rats acutely exposed to hypoxia.   These results indicate that normobaric hypoxia at 10.5% O_2 or 7.6% O_2 lowers the digestibility, which appears to recover with elongating the period of hypoxia.   Therefore, under hypoxic conditions, one possible cause for reduced food intake may be the decreased digestibility.   Further, because the decreased digestibility lowers the absorption of diet, weight loss due to hypoxia may   be attributable to hypoxia-suppressed digestibility.

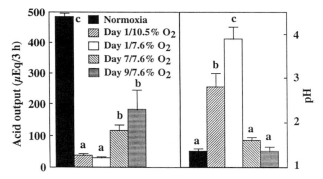

Figure 3. Effect of hypoxia on gastric acid secretion in conscious rats with pylorus ligation.   Conscious rats with pylorus ligation were exposed to hypoxia for 3 h and gastric acid was measured.   Values are expressed as means ± SEM of 10 rats/group.   Values with different letters superscripts are significantly different (P < 0.05).

The main stimulant of gastric acid secretion is gastrin, which is released from G cells in the glands of the gastric antral mucosa in response to cholinergic and peptidergic stimulation, and intragastric luminal peptides, amino acids and other nutrients. When the fasted rats are exposed to 10.5% or 7.6% O_2 hypoxia, hypoxia results in the rise in the gastric pH and elevates the plasma gastrin concentration (18). However, when rats with pylorus ligation are orally administered HCl and exposed to 7.6% O_2 hypoxia, the rise in the plasma gastrin concentration is not observed (18). These results demonstrate that the inhibitory effect of normobaric hypoxia on gastric acid secretion stimulates the gastrin release through positive feedback regulation.

Literature Cited

1. Boyer, S. J.; Blume, D. J. *Appl. Physiol.* **1984,** *57*, 1580-1585.
2. Guilland, J. C.; Klepping, J. *Eur. J. Appl. Physiol. Occup. Physiol.* **1973**, *54*, 517-523.
3. Rose, M. S.; Houston, C. S.; Fulco, C. S.; Coates, G.; Sutton, J. R.; Cymerman, A. *J. Appl. Physiol.* **1988**, *65*, 2545-2551.
4. Sridharan, K.; Malhotra, M. S.; Upadhayay, T. N.; Grover, S. K.; Dua, G. L. *Eur. J. Appl. Physiol. Occup. Physiol.* **1982**, *50*, 148-154.
5. Gill, M. B.; Pugh, G. C. E. *J. Appl. Physiol.* **1964**, *19*, 949-954.
6. Grover, R. F. *J. Appl. Physiol.* **1963**, *18*, 909-912.
7. Hannon, J. P.; Sudman, D. M. *J. Appl. Physiol.* **1973**, *34*, 471-477.
8. Kellog, R. H.; Pace, N.; Archibald, E. R.; Vaugh, B. E. *J. Appl. Physiol.* **1957**, *11*, 65-71.
9. Consolazio, C. F.; Johnson, H. L.; Krzywicki, H. J.; Daws, T. A. *Am. J. Clin. Nutr.* **1972,** *25*, 23-29.
10. Pugh, L. G. C. E. *Br. Med. J.* **1962,** *2*, 621-627.
11. Butterfield, G. E.; Gates, J.; Fleming, S.; Brooks, G. A.; Sutton, J. R.; Reeves, J. T. *J. Appl. Physiol.* **1992**, *72*, 1741-1748.
12. Consolazio, F. F.; Matoush, L. O.; Johnson, H. L.; Daw, T. A. *Am. J. Clin. Nutr.* **1968,** *21*, 154-161.
13. Highman, B.; Altland, P. D. *Arch. Path.* **1949,** *48*, 503-515.
14. Imray, C. H. I.; Chesner, I.; Winterborn, M.; Wright, A.; Neoptolemeus, J.-P.; Bradwell, A. R. *Int. J. Sports Med.* **1992**, *16*, 87 (abs.)
15. Kayser, B.; Acheson, K.; Decombaz, J.; Fern, E.; Cerretelli, P. *J. Appl. Physiol.* **1992**, *73*, 2425-2431.
16. Rai, R. M.; Malhotra, M. S.; Dimri, G. P.; Sampathkumar, T. *Am. J. Clin. Nutr.* **1975**, *28*, 242-245.
17. Milledge, J. S. *Br. Med. J.* **1972**, *3*, 557-558.
18. Yamaji, R.; Sakamoto, M.; Miyatake, K.; Nakano, Y. *J. Nutr.* **1996**, *126*, 673-680.
19. Bromster, D. *Scand. J. Gastroenterol.* **1969**, *4*, 193-201.

Chapter 25

Photon Emission by Heme-Protein in the Presence of Reactive Oxygen Species and Phenolic Compounds

Y. Yoshiki[1], M. Kawane[1], T. Iida[1], H. Yuan[1], Okubo[1,3], T. Ishizawa[2], and
S. Kawabata[2]

[1]Department of Environmental Bioremediation, Graduate School of Agriculture,
Tohoku University, 1-1 Amamiyamachi, Tsutsumidori, Aoba-ku, Sendai 980, Japan
[2]Tohoku Electronic Industrial Company, Ltd., Rifu Office, 6-6-6 Shirakashidai,
Rifu-Cho, Miyagi 981-01, Japan

The addition of H_2O_2 to cytochrome c, SOD and HRP gave a short
burst of chemiluminecence (CL). These CL increased the intensity
in the presence of phenolic compounds such as catechins.
Especially, gallic acid showed remarkable enhancement of CL
intensity. CL spectra are not responsible to any compounds
combined for CL, because of 660 nm, 610 nm and 520 nm for the
H_2O_2/gallic acid/MeCHO, H_2O_2/epigallocatechin/MeCHO and
H_2O_2/gallic acid/HRP system, respectively. The results suggested
that the CL arose from the complex of phenolic compounds and
MeCHO or HRP. The evidence of the ESR spin trapping method
showed that the hydroxyl radical (HO•) inhibition of cytochrome c,
SOD and HRP increased dramatically by the addition of gallic acid.
One mM gallic acid and 1.1 μM HRP mixture showed complete
inhibition of HO•, despite 21% inhibition on calculation. They
indicate that CL occur when reactive oxygen species are scavenged.
Furthermore, autoxidation of the linoleic acid study indicated that
the complex of gallic acid and MeCHO, cytochrome c, SOD or
HRP played as electron donor, although MeCHO, cytochrome c,
SOD and HRP had hydrogen abstractive activity.

We have found, previously, that photon emission was exhibited in the presence of
reactive oxygen species (X), catalytic species (Y) and receptive species (Z) at room
temperature and at neutral conditions. Low-level chemiluminescence (CL) of
phenolic compounds as Y was demonstrated in the presence of reactive oxygen and
acetaldehyde which acted as the typical Z. The occurrence and intensity of their
photon emission were closely related to the particle structures of the phenolic
compounds and the radical scavenging activity (Yoshiki et al., 1995a; 1995b).
Further, we have shown that the CL of catechin on the XYZ system was given by
equation [P] = k[X][Y][Z] (P, photon intensity; k, photon constant; [X], reactive
oxygen species concentration; [Y], catalytic species concentration; [Z], receptive
species concentration (Yoshiki et al., 1996). The CL in this system is thought to

arise from electron translation or hydrogen abstractions between S, Y and Z, because properties of Y are that of a radical scavenger and an electron donor like phenolic compounds.

In organic systems, various enzymes, especially heme-protein such as peroxidase, and cytochrome c are associated with reactive oxygen metabolism and/or electron translation. A previous investigation of ours indicated that horseradish peroxidase (HRP), which decomposes hydrogen peroxide to oxygen and water in consecutive one-electron steps, acted as Z in this CL system.

The present communication reports the results bearing on the generation of CL associated with the heme-pretein and on the effects of hydroxyl radical scavenging activity based on the XYZ system. It is believed that comparison of CL data with reactive oxygen scavenging activity on the XYZ system can lead to a better understanding of the pathways involved in the process.

Materials and Methods

Reagent. Gallic acid and linoleic acid purchased from Nakarai Tesque Co. (Japan). Hydrogen peroxide (H_2O_2) 30% was from Santoku chem. (Japan and acetaldehyde (MeCHO) was from Merck (Darmstadt), respectively. Horseradish peroxidase (HRP) and cytochrome c were from Sigma (St. Louis), and superoxide dismutase (SOD) from Nakarai Tesque Co. (Japan). 5,5-Dimethyl-1-pyrroline-N-oxide (nitrone spin tra) was from Mitsui Toatsu Chem. (Japan).

Chemiluminescence Measurements. The chemiluminescence of H_2O_2/catechins/MeCHO, cytochrome c, SOD or HRP were measured using a CL Detector single photon counting method, and was connected to a Waters Model 510 pump and a U6K injector. The wavelength was set in the range of 300-650 nm. The photons were counted as total intensities. The mobile phase used was 50 nM phosphate buffer (pH, 7.0); a flow rate of 1 mL/min and a temperature of 23 °C were used according to the methods described previously (Yoshiki et al., 1995a; 1995b; 1996). Ten micro liters of H_2O_2, gallic acid and MeCHO, cytochrome c, SOD or HRP were directly injected into the sample injector (final volume, 30 µL).

Determination of Emission Spectra. The spectrum of chemiluminescence was measured by a Simultaneous Multiwavelength Analyzer Model CLA-SP2 (Tohoku Electronic Industrial Co., Ltd.) equipped with a deffraction grating built in a spectroscope and a two-dimensional photomutiplier tube. The wavelength range of the spectroscope was 400-800 nm. The reaction mixture contained 50 mM phosphate buffer (pH, 7.0), 0.4 mL of 0.18 M H_2O_2, 5 mM gallic acid or 9.6 mM (-)-epigallocatechin, and 0.35 M MeCHO or 1.74 µM HRP in a quartz cell (final volume 1.2 mL). The light emission was determined for 180 sec.

Measurement of Hydroxy Radical Scavenging Activity. The ESR spectra were recorded on a JEOL JES-RE1X spectrometer using an aqueous quartz flat cell (60 mm X 10 mm X 0.31 mm, effective volume 160 µL). The optical absorption spectra were measured with an Otsuka Electronics MCP-100 multi-channel photodetector. Hydroxyl radicals (HO•) were generated in a system consisting of $FeCl_2$ and H_2O_2 (Fenton Reaction). The inhibitory ratio was calculated from the DMPO-OH signal as described by Mitsuta et al. (1990).

Measurement of Antioxidative Activity. Different amounts of samples were dissolved in 100 μL of 50% MeOH that were added to a mixture of 1.026 mL of 99% ethanol containing 2.51% linoleic acid and 2 mL of 50 mM phosphate buffer (pH, 7.0). the total volume was made up 5 mL by addition of distilled water. One hundred microliter of a mixture was used as an antioxidative assay with the ferric thiocyanate method as described by Inatani et al. (1983).

Results and Discussion

The addition of H_2O_2 to cytochrome c (cyt c), SOD and HRP gave a short burst of chemiluminescence (CL). Recently, we have shown through the CL method that gallic acid and MeCHO is typical Y and Z, respectively, on the XYZ system. On the basis of on this system, cyt c, SOD and HRP as Y or Z activity were detected by combining it with gallic acid or MeCHO through CL method. Although equal CL intensities to the H_2O_2/cyt c, SOD or HRP system were observed from H_2O_2/cyt c, SOD or HRP/MeCHO system (approximately 30 counts/sec), the H_2O_2/gallic acid/cyt c, SOD or HRP system gave high intensities (approximately 10^2-10^3 counts/sec) The intensity of CL suggested that the cyt c, SOD and HRP acted as the Z on the XYZ system. Therefore, the photon intensities of cyt c, SOD and HRP for Y or Z activity were compared with that of reactive oxygen species (X), 5 mM gallic acid (typical Y) and 0.35 M MeCHO (typical Z) mixture (control). The H_2O_2 or HO•/cyt c, SOD or HRP/MeCHO and H_2O_2 or HO•/gallic acid cyt c, SOD or HRP system were used to determine the activity as a Y and Z. Photon intensity as Z for cyt c, SOD and HRP showed 17, 5 and 277 fold increase in the presence of H_2O_2, and 19, 18 and 36 fold increase in the presence of HO•, in comparison with that of Y (Table I). The results indicated clearly that the cyt c, SOD and HRP acted as Z on the XYZ system. The heating HRP and SOD decreased and the maximal CL intensity dependent upon the heating time. However, in the case of cyt c, the strong CL was observed after treatment with heating. This difference could be partly explained by the steric structure between HRP and SOD which have a fifth ligand and a cyt c which has six ligand. Since the apo-protein obtained by treating HRP with acid-butanone, as described by Yonetani (1967), did not emit CL, it was thought that heme iron and peripheral amino residues played an important role in the CL of the heme-protein in the XYZ system.

It is known that phenyl compounds such as aromatic amino acids and phenols act as electron donors. In order to assess the role of Y on the intensity of CL, a comparison was carried out employing eight catechin compounds and a gallic acid. CL intensities were measured in the H/catechin/MeCHO system, the CL intensity was affected remarkably by the following catechin structures: a) the steric structure between the C-3 hydroxyl group and the B-ring, which is epimeric with respect to carbon 2, and b) a pyrogallo, rather than a catechol, structure in the B-ring. Although a similar tendency was observed in the H_2O_2/catechin/HRP system, the differences in intensities were only slight in comparison with that of the H_2O_2/catechin/HRP systems despite having the lowest intensity in the MeCHO system (Table II). The different intensities of both systems could indicate the importance of a Y peculiarities, *i.e.* molecular size, substrates, stereospecificity in interacting with the active site of Z, and especially high molecular weight compounds such as hem-protein.

Table I. Photon intensity of cytochrome c, SOD and HRP as receptive species (Z) in the presence of reactive oxygen species (X) and gallic acid as catalytic species (Y)

	H_2O_2		HO•	
	Y	Z	Y	Z
control	100	100	100	100
77 µM cyt c	21	357	4	76
77 µM SOD	9	44	5	88
77 µM HRP	22	6100	13	466

The photons of reactive oxygen (0.18 M H_2O_2, 0.025 M HO•), 5 mM gallic acid (typical Y) and 0.35 M MeCHO (typical Z were arbitrarily set at 100 units (control). Photon intensity for activity as Y of cyt c, SOD and horseradish peroxidase (HRP) was measured with reactive oxygen species (X) and 0.35 M MeCHO (Z). Photon intensity for activity as Z of cyt c, SOD and HRP was measured with X and 5 mM gallic acid.

Table II. Photon intensity of catechins or gallic acid as Y in the presence of hydrogen peroxide as X and actaldehyde or HRP as Z

	MeCHO	HRP
(-)-Catechin	1.3×10^3	1.0×10^3
(-)-Epicatechin	52.7×10^3	0.6×10^3
(-)-Gallocatechin	22.6×10^3	2.2×10^3
(-)-Epigallocatechin	4640.0×10^3	7.0×10^3
(-)-Catechin gallate	1.3×10^3	1.9×10^3
(-)-Epicatechin gallate	3.2×10^3	0.7×10^3
(-)-Gallocatechin gallate	6.8×10^3	4.6×10^3
(-)-Epigallocatechin gallate	356.6×10^3	4.9×10^3
Gallic acid	0.8×10^3	261.5×10^3

Assay conditions; 0.18 M H_2O_2 10 µL, 5 mM catechin or gallic acid 10 µL, and 0.35 M MeCHO 5 µL or 1.76 µM HRP 5 µL. Photon intensity is give by CPS (counts/sec) units.

The spectral analysis of the CL showed a distribution entirely in the visible region with a main peak at 660 nm, 610 nm and 520 nm for the H/gallic acid/MeCHO, H_2O_2/epigallocatechin/MeCHO, and H_2O_2/gallic acid/HRP systems, respectively (Fig. 1). These emission spectra, actually recorded during XYZ system, were distinguishable from that of singlet oxygen molecules as observed by Slawlnska and Slawlnski (1975) (Emax: 643, 702, 762, 1070, 1270 nm). Earlier evidence suggested that the singlet oxygen molecules were the main products which caused CL in heme-protein and contained enzyme like a peroxidase (Kanofsky, 1983). The spectral analysis, however, indicated a predominance of excitable carbonyl compounds in the XYZ system. Furthermore, spectral analysis indicated that the Y and Z complex caused the CL on the XYZ system, since spectra were corresponding in neither the Y nor the Z compounds.

To define biological or physiological the role of the XYZ system, hydroxyl radical (HO•) scavenging activities were measured by the ESR spin trapping method. The HO• scavenging activities of 70 µM cyt c, 32 mM SOD, 22 µM HRP and 1 mM gallic acid were 31.1%, 6.0%, 34.2% and 10.3%, respectively. The addition of gallic acid to the cyt c, SOD and HRP showed the dramatically synergistic inhibition of H (Table III). Especially, the effect was remarkable with the HRP. Although the HO• inhibition of 1 mM gallic acid and 1.1 µM HRP mixture is calculated at 21%, the mixture showed almost 100% inhibition. These results obtained by the ESR spin trapping method suggest that the photon emission on the XYZ system arise from energy translation when reactive oxygen species are scavenged by Y and Z compounds.

Table III. Effects of the cytochrome c, SOD, HRP and gallic acid on hydroxyl radical scavenging activity

	without gallic acid	with 1 mM gallic acid
70.0 µM cyt c	31.1 (%)	78.3 (%)
7.0 µM cyt c	5.6	26.5
32.0 mM SOD	6.0	29.3
32.0 µM SOD	0.1	10.9
22.0 µM HRP	34.2	97.9
1.1 µM HRP	10.7	95.2
1.0 mM gallic acid	10.3	–

cytc, cytochrome c: HRP, horseradish peroxidase

Physiological antioxidation protection has been used as one of the major defense mechanism in fighting free radical induced disorder. Therefore, autoxidation of the linoleic acid system is used to investigate the role of the Y or Z compounds, and the Y and Z mixture.

When linoleic acid was incubated with Z (MeCHO, cyt c, SOD and HRP), pro-oxidative activity was observed to be dependent on the Z concentrations. Gallic acid, which was generally thought to be a radical scavenger, acted as an antioxidant. Interesting activity was observed in the linoleic acid/gallic acid/Z system, that is

(A) H₂O₂/gallic acid/MeCHO system

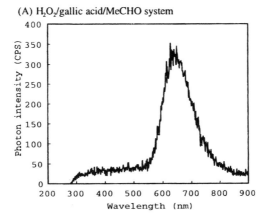

(B) H₂O₂/epigallocatechin/MeCHO system

(C) H₂O₂/gallic acid/HRP system

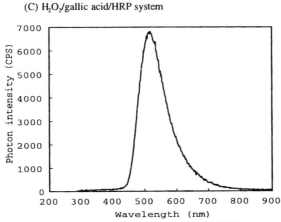

Fig. 1. Emission spectra observed in H_2O_2/gallic acid/MeCHO, H_2O_2/epigallocatechin/MeCHO, and H_2O_2/gallic acid HRP systems. Assay condition are described in "Materials and Method" section. Photon intensity is given by CPS (counts/sec) units.

linoleic acid/gallic acid/Z system prolong the induction period of lipid autoxidation more than the linoleic acid/gallic acid system (Fig. 2). The inhibition of hydrogen abstraction or the scavenging of linoleic hydroeroxide must be responsible for the strong antiosidative activity of the linoleic hydroperoxide must be responsible for the strong antioxidative activity of the linoleic acid/gallic acid/Z system. If antioxidative activity is cause by the scavenging activity of the linoleic hydroperoxide, gallic acid//Z system was examined to explore their role by HPLC analysis (cosmosil ODS column, 4.6 X 250 mm, Nakarai Tesqu.) using MeCN-water-HOAc (6500 : 3500 : 3) as a mobile phase (flow rate; 0.9 mL/min). HPLC evidence suggested that antioxidative activity, caused by the inhibition of hydrogen abstraction from linoleic acid because linoleic hydroperoxide generated by autoxidation, increased linearly with reaction time (data not shown).

The reaction mixture contains linoleic acid (2.51%) without (control) or with the indicated additives (1 mM gallic acid or 0.1 mM gallic acid as Y and 10 mM MeCHO, 0.4 mg/mL cyt c, 0.02 µg/mL SOD or 13.9 nM HRP as Z).

Figure 3 shows possible mechanisms for the XYZ system during the autoxidation of linoleic acid. It is well known that the oxygen radicals attack lipids and abstract doubly allylic hydrogen from polyunsaturated lipids to initiate free radical-mediated chain oxidation. Z(MeCHO, cyt c, SOD, and HRP) has hydrogen abstractive activity in the autoxidation system of linoleic acid, since Z acts as a pro-oxidant (route 1). The synergistic antioxidative activity of the Y (gallic acid) and Z mixture could be inferred by route 2, that is by linoleic acid radicals receiving hydrogen generated by hydrogen- or electron-translation between Y and Z (route 3). The oxidation of heme-protein by H_2O_2 is currently explained interms of a heterolytic cleavage of O-O bond of the coordinated peroxide:

$$HX\text{-}Fe^{III} + H_2O_2 \longrightarrow \bullet X\text{-}Fe^{IV} = O + H_2O$$

where HX- stands for an amino acid in the protein and -Fe for the heme iron. Although oxidation state (Fe^{IV} = O) confirmed by the NMR technique and by calculation of spin densities, almost all of the experiment demonstrated the direct reaction of H_2O_2 and the heme-protein (La Mar et al., 1983; Loew et al., 1980). the CL properties such as CL intensity or spectra strongly supported the formation of the Y (gallic acid) and the Z (MeCHO or the heme-proten) complex.

Low-level CL in the heme-protein included in the enzymes have remained elusive due to an inability to reliably assess its quantitative production. In this study, we have examined the CL of the XYZ system to define the pathways involved in the process. CL in this system might primarily be scavenging mechanism of the reactive oxygen species. Inhibition or scavenger of the reactive oxygen species in the XYZ system might proceed via a hydrogen- or electron-transfer mechanism. It supports the synergistic effects on HO•, and causes inhibition or antioxidative activity in XYZ system.

Literature Cited

Inatani, R., N. Nakatani and H. Fuwa. Antioxidative effect of the constituents of rosemary (*Rosmarinus officinalis* L.) and their derivatives. *Agic. Biol. Chem.* **1983**, *47*, 521-528.

Kanofsky, R. J. Singlet oxygen production by lactoperoxidase. *J. Biol. Chem.* **1983**, *258*, 5991-5993.

(A) MeCHO (1 mM gallic acid, 10 mM MeCHO)

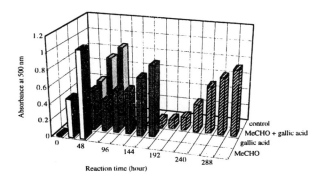

(B) cytochrome c (0.1 mM gallic acid, 0.4 μ g/ml cyt c)

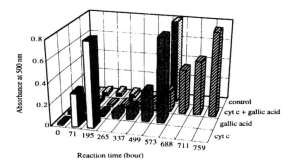

(C) SOD (0.1 mM gallic acid, 0.02 μ g/ml SOD)

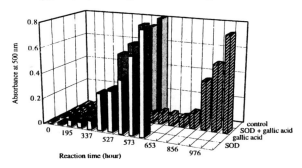

Fig. 2. Effects of gallic acid (Y) and receptive species (Z) on lipid autoxidation.
The reaction mixture containes linoleic acid (2.51%) without (control) or
with the indicated additives (1 mM gallic acid or 0.1 mM gallic acid as Y
and 10 mM MeCHO, 0.4 µg/mL cyt c, 0.02 µg/mL SOD or 13.9 nM HRP as
Z)

Continued on next page.

(D) HRP (1 mM gallic acid, 13.9 nM HRP)

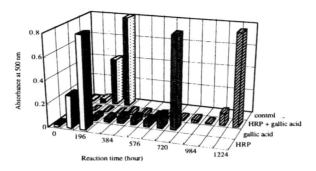

Figure 2. *Continued.*

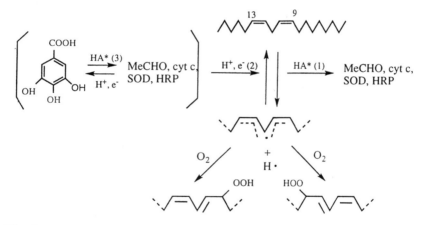

Fig. 3. Possible mechanism for XYZ system during autoxidation of linoleic acid.
*hydrogen abstraction

La Mar, N. G., J. S. de Ropp, L. Latos-Grazynski, A. L. Balch, R. B. Johnson, K. M. Smith, D. W. Parish and R. Cheng. Proton NMR Characterization of the ferryl group in model heme complexes and hemoproteins; Evidence for the $Fe^{IV} = O$ group inferry myoglobin and compound II of horseradish peroxidase. *J. Am. Chem. Soc.* **1983,** *105,* 782-787.

Loew, H. G., Z. S. Herman. Calculated spin densities and quadrupole splitting for model horseradish peroxidase compound I: Evidence for iron (IV) porphyrin (S-1) cation radical electronic structure. *J. Am. Chem. Soc.* **1980,** *102,* 6173–6174.

Mitsuta, K., Y. Mizuta, M. Kohno, M. Hiramatsu and A. The application of ESR spin-trapping technique to the evaluation of SOD-like activity of biological substances. *Bull. Chem. Soc. Jpn.* **1990,** *63,* 187–1991.

Slawlnska, D. and J. Slawlnski. Chemiluminescent flow method for determination of formaldehyde. *Anal. Chem.* **1975,** *47,* 2102-2109.

Yonetani, T. Studies on cytochrome c peroxides. *J. Biol. Chem.* **1967,** *242,* 5008–5013.

Yoshiki, Y., K. Okubo, M. Onuma and K. Igarashi. Chemiluminescence of benzoic and cinnamic acids, and flavonoids in the presence of aldehyde and hydrogen peroxide or hydroxyl radical by Fenton reaction. *Phytochem.* **1995a,** *39,* 225–229.

Yoshiki, Y., K. Okubo, and K. Igarashi. Chemiluminescence of anthocyanins in the presence of acetaldehyde and *ter*-butyl hydroperoxide. *J. Biolumin. Chemilumin.* **1995b,** *10,* 335–338.

Yoshiki, Y., T. Kahara, K. Okubo, K. Igarashi, and K. Yotsuhashi. Mechanism of catechin chemiluminescence in the presence of active oxygen. *J. Biolumin. Chemilumin.* **1996,** *11,* 131–136.

INDEXES

Author Index

Subject Index

An example of the application of the Genox OSP is where oxidative stress profiles were recorded in human subjects who were subjected to severe exercise. The levels of the lipid oxidation products, such as the total alkenals, lipid hydroperoxides, and thiobarbituric acid reactive substances (TBARS), in serum and urine were measured. It was observed that the levels of TBARS and total alkenals were far higher in the case of test individuals who were subjected to severe exercise than the control subjects. Further, it was also observed that the levels of these lipid oxidation products in the urine had a positive correlation with the levels of the oxidative DNA damage product, namely the 8-hydroxy-2'-deoxyguanosine (8-OHdG) (Figure 4).

Studies were also undertaken to demonstrate the effect of consumption of antioxidant green tea, developed by the Japan Institute for the Control of Aging, on the levels of 8-OHdG excreted in the urine. The study was carried out on 48 individuals, 31 males and 17 females, for a total period of 32 weeks. These individuals were divided into five age groups - those in their twenties, thirties, forties, fifties, sixty and above. These individuals were asked to drink five cups of the antioxidant green tea (1 g instant green tea dispersed per cup of 100 ml hot water) per day. The levels of urinary 8-OHdG in these individuals were tested daily during the 10 week control period, 12 weeks of test period when green tea was administered, and 10 weeks of post test control period. As far as possible the life style and the dietary habits during the entire period of the 32-week study of these subjects were kept under strict control.

Urine samples were suitably diluted for quantification of 8-OHdG by the 8-OHdG ELISA Assay Kit , developed by the Japan Institute for the Control of Aging (15). It was observed that the amount of the 8-OHdG excreted by the entire population, during the control period, ranged from 12.0 to 24.3 μg, whereas the corresponding levels in the same individuals dropped to levels ranging from 8.1 to 19.9 μg during the green tea consumption period (Table I). Further, it was also observed that the levels of 8-OHdG in the female subjects were found to be lower than the male subjects during both the control and the test period (Table I). These results indicate that the endogenous antioxidants present in the green tea offer protection against oxidative damage thereby lowering the oxidative stress state of these individuals.

Conclusion

Many individuals, from infants to young adults, show remarkable differences in their OSP. This difference is a result of dietary habits and general life style, and due to inherent genetic components that regulate the OSS. There is now sufficient evidence indicating that many people who appear to be healthy have a much higher than average OSS and are likely to be on an accelerated path developing an oxidative stress-related disease, such as cancer, diabetes or cardiovascular disease. There is, therefore, a need to determine an individual's OSP as early as possible in life so that corrective action can be taken, if necessary. Furthermore, an individual's OSP should be continually monitored throughout their life to determine how

193

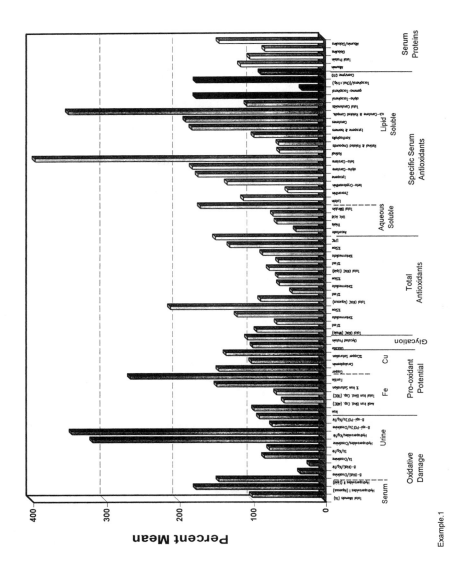

Figure 3. A typical example of the Genox Oxidative Stress Profile.

Example.1